Springer Series in Reliability Engineering

Series Editor

Hoang Pham, Department of Industrial and Systems Engineering, Rutgers University, Piscataway, NJ, USA

More information about this series at https://link.springer.com/bookseries/6917

Adolfo Crespo Márquez

Digital Maintenance Management

Guiding Digital Transformation in Maintenance

Springer

Adolfo Crespo Márquez
Department of Industrial Management
University of Seville
Seville, Spain

ISSN 1614-7839 ISSN 2196-999X (electronic)
Springer Series in Reliability Engineering
ISBN 978-3-030-97659-0 ISBN 978-3-030-97660-6 (eBook)
https://doi.org/10.1007/978-3-030-97660-6

This Springer imprint is published by the registered company Springer Nature Switzerland AG
The registered company address is: Gewerbestrasse 11, 6330 Cham, Switzerland

I dedicate this book to Lourdes del Castillo Silva, my love and life companion, our family's shelter and home.

Foreword

Asset Management is the set of coordinated activities that an organization performs in order to realize value from its assets. As a concept, asset management has gained prominence across all industry sectors, particularly since the publication of the ISO 5500x family of standards in 2014. A core component of asset management is maintenance management, which includes all the activities involved in the setting of maintenance objectives, strategies and responsibilities, and implementing them through proper planning, control and supervision aligned with the economic and sustainability aspects of the organization.

Although maintenance management and asset management are technically different, they are still interrelated and should work in an integrated manner. Maintenance management ensures that the physical performance of maintenance activities is carried out effectively and efficiently, while asset management encompasses all relevant aspects to ensure that the (properly maintained) asset delivers the greatest value to the business.

The past decade has also seen the rapid advances in digital technologies, data analytics and artificial intelligence approaches that has the potential to radically transform the way we manage, and specifically, maintain assets. The convergence of sensing/data-capture solutions, increasing computational power at ever-shrinking costs, and advances in statistical techniques to drive machine learning under the umbrella of the Fourth Industrial Revolution have equipped us with tools that help generate a deeper understanding of how complex industrial systems behave and perform, thus enabling us to manage them better. In this context, we have also seen the emerging popularity of the concept of Digital Twins, which aim to replicate physical equipment and systems in the digital world through effective integration of data, models and decision-support systems promising a step change in productivity and sustainable performance.

Despite encouraging developments in digital solutions, a number of challenges remain to be addressed before the potential opportunities they present can be realized effectively for maintenance. The challenges include dealing with the inherent uncertainties in complex systems and the environment they operate in, data quality problems arising from poor data management and governance practices, and a general lack of awareness of which techniques and technologies are suitable to tackle particular maintenance management problems.

It is under this backdrop of opportunities and challenges that this book offers interesting, innovative, and practical insights and guidance on digital maintenance management. Divided into six parts, the book includes twelve chapters that deliver a concise illustration of the key tools and techniques relevant to the development and application of digital approaches for asset management. Starting with an introduction and review of related concepts, the book offers a framework for data management, an overview of the tools currently available in the market, and the latest advances in research that can be used to address some of the challenges we face in managing the costs, risks and performance of our assets through systematic application of data-centric techniques. The heart of the book lies in the rich set of real-world industrial case studies that brings theory into practice. The book therefore constitutes an invaluable resource for not only academics and researchers, but also to industrial practitioners.

Ajith Kumar Parlikad, Ph.D., FISEAM, MIEEE, MIET
Professor of Asset Management
University of Cambridge
Cambridge, UK

Scientific Secretary
IFAC TC 5.1 Working Group on Advanced Maintenance
Engineering Services and Technology, University of Cambridge
Cambridge, UK

Former Chair
Academic and Research Network, Institute of Asset Management
Bristol, UK

Acknowledgements

The author wishes to thank specific people and institutions for providing their help during the years 2020 and 2021, making the publication of this book possible.

Pedro Moreu de León and Antonio Sola Rosique, the "seniors" and "wise men" of Ingeman, provided their reflections, advice and recommendations when it came to approaching each of the topics addressed in this work. These are highly technical and technological topics, and a complete, holistic view of each of them is necessary to fit them conveniently into a book such as the one now presented.

Juan Francisco Gómez Fernández co-authored several of the articles that serve as the starting point for this book. Thanks to his advice, many of the topics reviewed here are technically well focused as well as applicable in practice. Juan is an expert in processes, information systems and artificial intelligence tools for asset management.

Antonio Jesús Guillén López is a colleague of the SIM Research Group, his expertise in the asset data model to be used with CBM/PHM technologies, besides his work on a dynamic vision of risk and its relationship with maintenance actions, are important for the related chapters in this book.

Antonio de la Fuente Carmona is a young and experienced Ingeman engineer. Deeply knowledgeable about asset management and engineering methodologies, he has also delved into complex data science processes. His skills with the RapidMiner tool, coupled with his knowledge of AHI and CBM techniques and models, have made many of the results presented in the book possible. Antonio expects to finish his Ph.D. during the current year.

Jesús Ferrero Bermejo and Fernando Olivencia work for Magtel company and Jesús has recently finished his Ph.D. within our research group. In some Chapters of this book ML techniques are used for maintenance and energy efficiency management in Magtel solar plants. This was a challenge that we could face together with the help of Rafael Pino from the Department of Statistics and Operations Research of the University of Cordoba.

Similarly, over the last few years, Javier Serra Parajes and Fernando Hidalgo Mompeán co-authored several papers related to maintenance management and maintenance and capital investments analysis, within the Oil & Gas sector, especially dealing with LNG equipment and plants. Javier is nearing the conclusion of his Ph.D.

work within our group, resulting from an intensive collaboration of the SIM group and Ingeman with Enagas Corporation. In some of this work, Gonzalo Cerruela-Garcia, from the Department of Computing and Numerical Analysis of the University of Córdoba, provided an extremely valuable contribution with computational work and helped in the analysis and interpretation of results.

José Antonio Marcos Alberca is the Head of the Smart Maintenance Engineering and New Projects Department of Patentes Talgo, and Javier Navarro Iglesias is the iMaintenance Manager, of the same company. They also co-authored important papers for this book. Jose Antonio is also doing his Ph.D. with the SIM group and has always offered the opportunity for a strong collaboration between Patentes Talgo, Ingeman and the University of Seville. This collaboration has turned out to be critical to our group's production and to this book.

Eduardo Candón Fernández and Pablo Martínez-Galán have been responsible for intensive work for Ingeman and SIM Group Projects related to asset health indexing and risk assessment, results of their work are also included in this book. Pablo is about to complete his Ph.D. within the SIM group.

The rest of the members of the SIM group and Ingeman, not mentioned here, was working hard in parallel, on other projects, during this time.

Sara Antomarioni, currently Graduate Research and Teaching Assistant at Università Politecnica delle Marche working with Prof. M. Bevilacqua, visited Seville from Nov 2018 to Feb 2019, and during that period she really did reinforce our group skills, which were used later for this work.

Adolfo Crespo del Castillo provided valuable elements of the literature review for the chapter related to the dynamic scheduling of maintenance. Adolfo is currently acting as a consultant for Ingeman whilst studying for his Ph.D. with Prof. A. K. Parlikad at the Institute for Manufacturing of Cambridge University.

Over the last two years, the funding of the following projects was very important for the publication of this book: Methodology for industrial application of smart maintenance solutions. Integration of Predictive Analytics and Machine Learning techniques in IoT platforms (CEI-19); Support for the presentation of European proposals in the field of AIM modeling for the digital transformation of asset management (EIN2019-103326); and Strategic level information models for the management of digital assets' life cycle (PP2019-13230). This funding was provided by the Government of the Community of Andalusia, the Spanish Ministry of Education and the University of Seville, respectively.

As usual, I would like to end these acknowledgements by mentioning my wonderful family: Lourdes, Lourdes Jr., Adolfo Jr. and Gonzalo. Always offering me their unconditional love and support, enabling this work to be accomplished.

To all of them, thanks.

Overview

The purpose of this book is to study the digital transformation of maintenance management in the context of industry and infrastructures. According to the reports from various institutions, maintenance is one of the business areas where this transformation is expected to be more significant.

To understand this historical opportunity for maintenance, this book discusses:

- Why maintenance can benefit from this transformation.
- What are the new technologies and tools which will have the highest potential impact in maintenance and why?
- How this transformation process can be ordered and conducted by understanding the management needs of the maintenance organization.
- The need for suitable data models for maintenance management.
- The emerging intelligent maintenance and asset management platforms. Their organization and new compelling functionalities.
- Scientific challenges in the adoption of advanced digital tools for maintenance.
- The emerging processes improving strategic and operational decision-making in maintenance.

This book argues that the basic principles of maintenance management will continue to be relevant in this new era and will help to properly guide the development of digitization plans for more advanced management. To this end, learnings from the book "The maintenance management framework. Models and methods for complex systems maintenance", will be used to understand where development can be more convenient for a given business environment.

Relevant Topics

The use of digital data processing is revolutionizing industrial and infrastructure maintenance in many ways:

From a strategic point of view, digital data coming from assets can be used to quantify the general health of complex assets identifying the end of their useful life. Health indices can be developed taking into consideration the operational data, field inspections and laboratory tests data, with the intention to produce a quantitative indicator which may be used as a tool to manage assets, to identify capital investment needs and maintenance programs. This may allow one to compare the health of similar equipment placed in different functional locations, to study possible premature deterioration and optimize operation plans and/or asset maintenance; to communicate more accurately with manufacturers/builders; to understand the behavior of assets of different manufacturers/builders in specific functional locations; and to support decision-making processes in future investments in assets, or in the analysis of their life's extension.

From a more operational point of view, sensors on assets critical components can provide enormous amounts of data points for the analytics to detect anomalies, ensuring maintenance is only done when required, before a defect occurs. A proper estimation of the remaining useful life of the components allows meeting very high reliability requirements for close to 100% availability, once corrective maintenance activities can be carried out when units are not currently in service, avoiding downtime.

A real breakthrough in digitalizing maintenance can be envisaged when a growing number of systems will be interlinked within industrial plants and infrastructures ecosystems. This may lead to more sophisticated scenarios where collaboration among assets can play a critical role. A principle of collaborative learning may help to gain intelligence in several areas of strategic and operational maintenance management.

Regardless of the business activity level where the data is used, the expected high volume of maintenance data will demand new business processes and IT systems. Computational needs will now be on "the edge" or in "the cloud" to increase efficiency of advanced Apps delivering new functionalities to the assets end users. The servitization of the products of traditional OEM and the development of new digital services, such as Predictive Maintenance as a Service (PMaaS) or Simulation as a Service (SaaS), are expected to be here to stay.

Structure of the Book

The book is divided into six Parts. In Part 1, Chapter 1 introduces the fundamental concepts explaining the benefits of digital transformation for maintenance management. Chapter 2 introduces new technologies that could be relevant to different dimensions of maintenance management.

In Part 2, the maintenance management framework is reformulated to benefit from digital transformation. Chapter 3 explains the transition to a model for Digital Maintenance Management (DM3). The elements of this model are introduced, and their functionality is explained. Chapter 4 is about data models. The concept of asset data model (ADM) is taken as a basis to drive day to day management decisions and activities in the digital world.

Part 3 of the book is dedicated to the study of advanced Apps for asset management and maintenance. In Chap. 5, different interesting tools are presented and classified as APM or AIP categories, according to their scope: asset performance management or asset investment planning, respectively. Some industrial vendors and IT vendor suites are analyzed. Their modules are listed and those related to the DM3 model are selected and detailed. The scalability of these emerging apps, is another topic discussed, trying to find a compromise between their functionality and the computational power that they provide.

Part 4 is devoted to new emerging challenges related to the introduction of these new technologies, topics such as ergodicity, dimensionality, the consideration of risk, and dynamic maintenance scheduling are discussed in Chaps. 6–9.

Parts 5 & 6 provide a set of good examples of emerging digital maintenance management processes, at different activity levels in the organization, showing the way maintenance and assets management in general, can benefit from a proper implementation of all these technologies and methodologies. Chapters 10 and 11 are related to asset performance improvement processes while Chap. 12 describes a methodology to support the process of asset investment planning.

Adolfo Crespo Márquez
Escuela Técnica Superior de Ingenieros de Sevilla
Isla de la Cartuja
Sevilla, España

Contents

Part I
Fundamental Concepts to Understand Digital Transformation Benefits

Chapter 1
Benefits of Digital Transformation for Maintenance Management Systems. Market Trends

1.1 Introduction

The digital transformation is changing the way maintenance is managed. The new maintenance strategies are reaching the objective of increasing assets dependability over the life cycle while reducing the frequency of scheduled maintenance. How is this possible? What is the process to follow, and what are the tools to use, to reach such excellence in management? How can we adapt our current maintenance management system to a digital management system?

Clearly, the process is not trivial neither is the tools implementation sequence. To develop a maintenance digital strategy, we must understand that the existing maintenance plans will have to be deeply analyzed and probably broken down for each specific system, often lowering the level of indenture of the functional locations, according to opportunities highlighted by new technologies and data analysis.

Methods like reliability centered maintenance (RCM) will play a pivotal role. This is a technique organizing the entire maintenance activities over critical assets, determining potential task to be realized, delayed, or eliminated using real-time monitoring and more advanced CBM strategies. RCM allows keeping track of changes in maintenance plans and can help to guide the course of action with suitable risk–benefit analyses.

Many manufacturers, asset users and managers must prepare for various possible scenarios according to the digital configuration of the asset, leading to different possible maintenance plans and management needs. For instance, predetermined maintenance task can be delayed or replaced by CBM in some, but not all projects. Preventive maintenance scheduled can be maintained, but the maintenance time can change according to some project pre-validated maintenance operations, using different technologies, etc.

Moreover, the new digital information flows require a change in the management processes to support decision-making. Business intelligence apps together with powerful visualization tools will have to be implemented to improve man-systems interaction, fleet maintenance scheduling and shop floor management.

A. Crespo Márquez, *Digital Maintenance Management*, Springer Series in Reliability Engineering, https://doi.org/10.1007/978-3-030-97660-6_1

As the reader may guess, one of the biggest challenges will be to provide proper change management support and training for people. While the younger generation are familiar with data and computer-based work tools, older employees may be unwilling to change the way they work and often need support. Best algorithms for detection, diagnosis and prognosis will serve for nothing if maintenance staff do not change the way they work.

At the same time, and to realize impact, the entire value chain in maintenance needs to be addressed; monitoring and predictive analytics represent only the first step. In the new maintenance ecosystem, critical attention must be paid to ownership of the data masters and possibilities that these masters offer for management. Building analytical capacities enabling a successful operating model may result in a matter of data access negotiation.

In summary, digitalization is going to change the way we conceive, develop, and industrialize maintenance. This must be understood as a long, complex, and formal process. A process that should not be implemented all at once but must be consolidated quietly and safely.

1.2 Key Managerial Areas That Can Be Digitally Supported

Most of the data that is collected through new digital technologies can be used to increase the effectiveness of asset management decision-making. Consequently, the types of asset management decisions and their importance in meeting the organization's AM objectives need to be identified.

The importance of each reporting requirement associated to a given decision-making process can be estimated by identifying the benefits of better a reporting and decision making and the risks associated with poor (non-data-driven) decision making or reporting. To this end, the process to follow could be [1]:

1. Identify the requirements of all key stakeholders with respect to the provision of information.
2. Identify the most critical assets and then select the information that must be considered that will lead to effective decision making. Finally, it is necessary to determine what data is needed to obtain this information. The necessary data could take many different forms, including:
    - Data about the assets (themselves, current condition, current level of performance).
    - Data relating to the activities that have been performed on the assets (operational activities, maintenance activities and modifications, upgrades or replacements);
    - Data about the financial or other impacts if the assets underperform or fail to perform at all.

- Data relating to safety, environmental or other incidents.
- Data relating to expected future asset performance, costs, and risks.
- Etc.

3. Identifying the types of decisions which will have the greatest potential impact on the achievement of asset management (and organizational) objectives. The decisions can be made at many different organizational levels, including:
   - *Strategic Decisions*—potentially those with the greatest potential business impact, capital investment and allocation of operating expenditure decisions. Also, those decisions for which objective data is most likely to be difficult to obtain and analyze.
   - *Management Decisions*—such as to replace or upgrade an asset to meet specific business needs, about the timing of these major events, also those related to the allocation of working capital (such as for spare parts holdings), decisions relating to whether to insource or outsource activities.
   - *Operational Decisions*—involved with short term control of maintenance and operational activities, these are technical decisions relating to day-to-day Operations.

Some experts agree that there has been unequal attention paid to the *management* and *operational* areas over the last decades where many business resources were spent on activities to get the jobs done more efficiently (right hand side of Fig. 1.1, a figure adapted from Asset Management (AM) decisions and controls of Woodhouse [2]). For instance, master asset registers were created, work management systems and procurement systems were developed, and communication with craftsmen was improved. However, it seems that a better balanced and more *strategic* view is needed (Left hand side of Fig. 1.1), understanding business priorities in work and investments. For instance, simple and robust methods for project evaluation and investment, or to consider life cycle extension prior to reinvestment, are still needed [3].

1.3 Emerging Asset Management Systems APM & AIP

Nowadays companies are complementing original Enterprise Asset Management Systems (EAM) or Computerized Maintenance Management System (CMMS), with different solutions/Apps to gain more capabilities in those managerial areas mentioned in the previous Section. What are these solutions? How can they be classified? What should be the most convenient to implement in each business? To answer these questions let us first take a look at what the market situation in Asset Management Systems is.

Concerning current market situation, Gartner's [4] distinguishes among three different types of asset management systems in the market: EAM, APM and AIP Solutions.

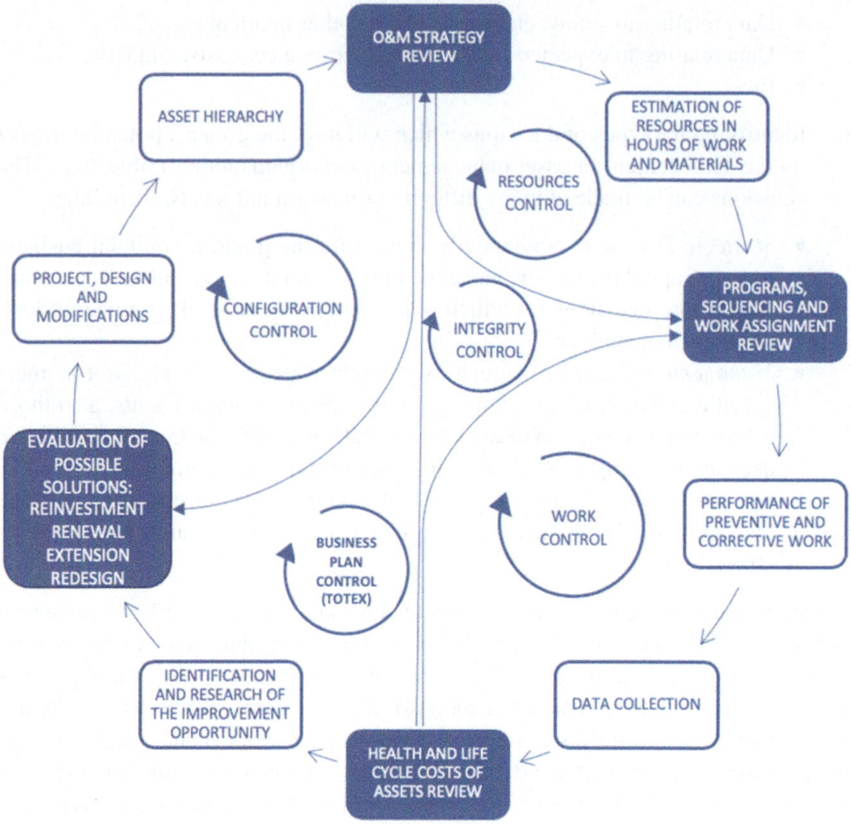

Fig. 1.1 A perspective of AM Decisions and controls

APM and AIP benefit from connectivity and digital technologies, often share the same data and use similar predictive analytics, but with different aims, as follows:

- Investments in EAM are made to manage asset inventory, configuration, and maintenance execution.
- Organizations are willing to invest in *APM tools* and technologies to reduce corrective maintenance, increase availability, and reduce the risk of failure (especially when assets are critical). Moreover, APM tools may increase an organization's ability to comply with regulations related to assets inspections and maintenance. Data capture, analytics and visualization, allow these tools to improve operations and maintenance timing, and to identify which maintenance and inspection activities to perform on industrial assets [4].
- The APM market has been around for more than a decade, but the vendor community has recently doubled its R&D [7], which is now boosting growth rates. The growing maturity of communications infrastructure at industrial sites

supports broader APM adoption as does a wave of new CEOs launching digital transformation strategies.

- Investments in AIP software are made to enhance complex, longer-term tactical and strategic decisions relating to the CAPEX/OPEX budget allocations and overall asset management planning [5]. AIP tools predict asset's current and future performance and link that expected performance to different investments options over a predefined mid- or long-term horizon. Investment planning for asset-heavy and process-intensive industries is now more relevant than ever. Economic recession has exacerbated this situation and executives are steadily leaning on in AIP software to enhance these decision-making processes.

The reader should not confuse APM with AIP, APM is designed to support safe, reliable, and efficient operation of equipment and infrastructure, while AIP is designed to support both short- and long-term capital investment decisions. See Verdantix vision of the three types of systems in Fig. 1.3 [5] (Fig. 1.2).

Some EAM vendors have an APM product strategy, but most rely on partnerships with APM vendors. It seems clear that sales of the APM Apps may depend on whether they are compatible with the existing EAM system.

AIP is used most heavily by local governments and regulated utilities, such as power, oil & gas but is also increasingly used in other industries, such as water, transportation facilities, and telecom, and is of interest to manufacturing and healthcare industries.

Curiously, market researchers have found very few overlaps between APM and AIP solution providers (see [5, 6]).

To conclude this Section, and according to previous Sections, different decision-making areas, as depicted in Fig. 1.1, are supported by different asset management

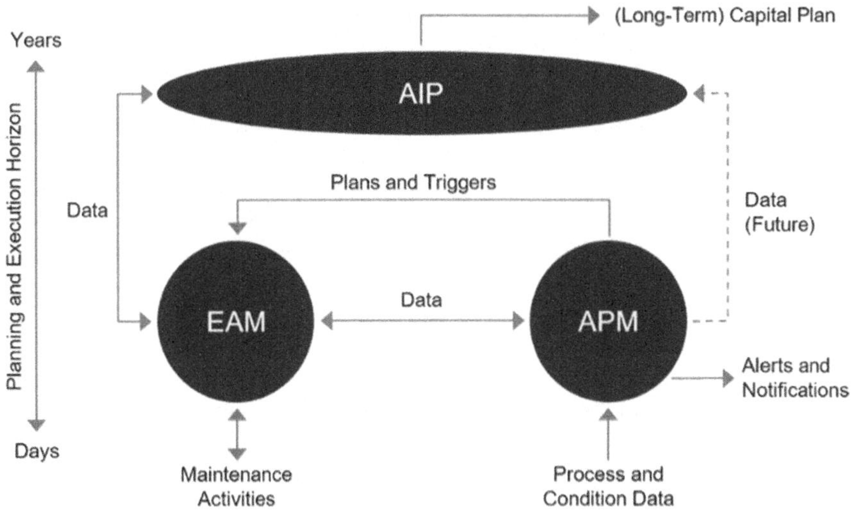

Fig. 1.2 The flow of data in advanced asset management systems

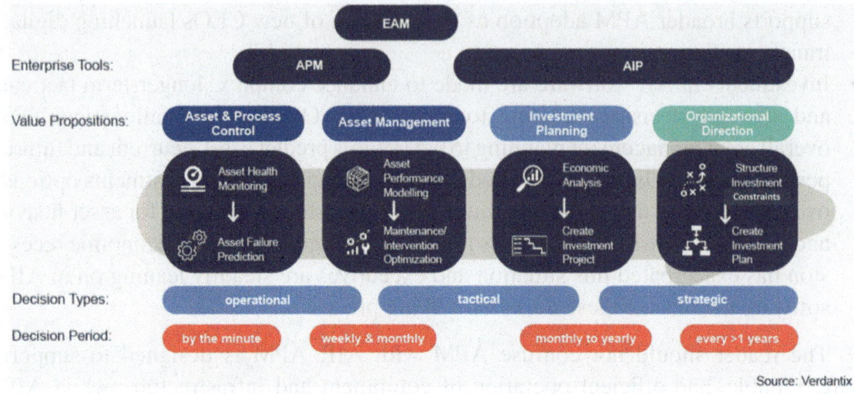

Fig. 1.3 Verdantix vision of EAM, APM & AIP software solutions [5]

systems, as explained in Fig. 1.2. A graphical explanation of how managerial areas are supporting by systems is illustrated in Fig. 1.4.

1.4 The New Am Systems/Platforms Architecture

Concerning the implementation of modern APM & AIP systems, companies can be placed in one of the following three basic scenarios:

- Some companies are assuming developed solutions from assets providers [7],
- Other companies are developing tailor-made solutions internally [8], and
- Other companies' developments are supported over big data intelligent solutions configured over cloud IT solutions [9].

The similarity in concepts application and compatibility with new assets or among systems is still very difficult to sustain and most of the time requires an enormous financial effort.

Relying on Cloud Computing definition [10], the reader may expect these systems to be implemented as cloud-based PaaS (Platform as a Service: a class of cloud computing services that supplies a platform supporting the development, running, and management of applications for asset management) providing a standard way to connect sensing elements, instruments, and other devices together with computers' industrial applications and people.

A main advantage of a PaaS environment is that software developers do not have to worry about some of the lower-level details of the environment. A software stack can be considered as a pyramid [11, 12].

As shown in Fig. 1.5, the pyramid is divided into the following three levels:

- Infrastructure as a service (IaaS): It is located at the lowest level of the pyramid and contains among other capabilities the operating systems, networks and storage.

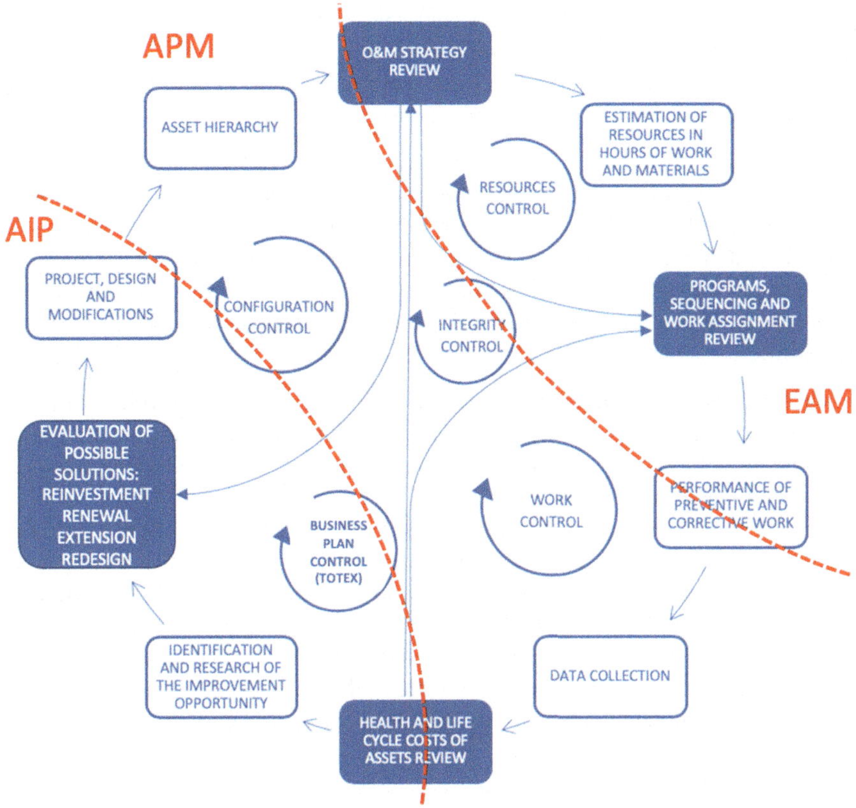

Fig. 1.4 Asset management decision areas and the EAM, APM & AIP Systems

Fig. 1.5 IAMP software stack pyramid

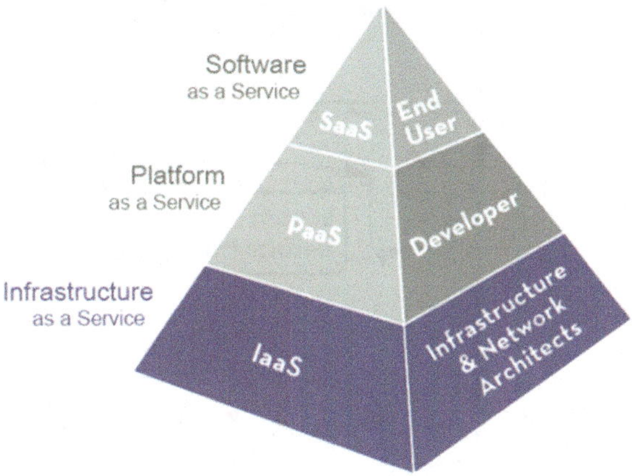

- Platform as a service (PaaS): It is located at the medium level and includes services related to the development and implementation of applications.
- Software as a service (SaaS): It is located at the top of the pyramid and represents the applications (Apps) provided to end users.

The Apps provided, by different industrial and IT vendors, to end users in the SaaS level of the pyramid are structured according to principles of software hierarchy design [13], as presented in Fig. 1.6. Normally, at present, an advanced asset management system or platform encompasses "everything needed" for "digital asset management". Examples of these names are GE's Predix, SAP's IAM or SIEMENS' MindSphere. Under the umbrella of this commercial name the software is structured as follows:

- On a first level, M number of modules can be found, each one referring to different domains within asset management. These modules are inter-dependent, in fact, their interrelation can be crucial. Examples of modules are reliability module, health module, strategy module, etc.
- On a second level, each module contains Apps that will have different purposes according to the scope of the specific decision-making process that they support. For instance, Apps will be found dealing with the hierarchy of assets, with the analysis of recurrent failures, with the design of the maintenance strategies, with the development of Condition Based Maintenance (CBM) programs, with the assessment of the assets' health, integrity and their life cycle cost, with the performance and analysis of the maintenance processes and activities, etc.

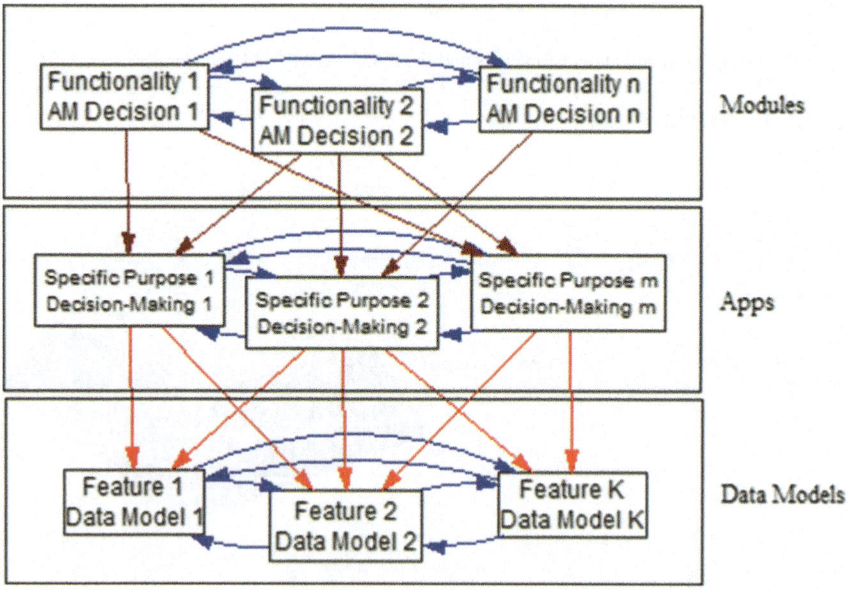

Fig. 1.6 Structure of Apps at the SaaS level

- Finally, on a third level, each application contains a data model. The data model serves to configure the databases where the relevant data for the Apps will be stored, with k-level scalability.

The Apps offered to end users by different vendors sit at the top of the software pyramid, and they benefit from the services provided by the platform (basically ETL, data storage and virtual machines services). The main challenge of these SaaS packages is to provide friendly user customization in a comprehensive way, to ensure each module's sustainably with minimum effort in data acquisition and manipulation. This is solved thanks to the framework described above, which links different data models, correlating data tables structurally, like human reasoning, to support the corresponding decision-making processes, without the need to know the complexity of the low-level software. Within this environment, appropriate asset data models are compulsory.

References

1. Dunn S (2019) Asset management data and decision making—implementing ISO 55000. 2019. https://www.assetivity.com.au/article/asset-management/asset-management-data-and-decision-making-implementing-iso-55000.html
2. Woodhouse J (2001) Asset management decision-making. http://plant-maintenance.com/articles/AMdecisions.pdf
3. Woodhouse J (2003) Asset management: a science emerging. Int J COMADEM
4. Gartner's Market guide for Asset Performance Management Software, dated 26th June 2019 (ID G00388410)
5. Voulgaropoulos (2021). https://www.verdantix.com/blog/asset-investment-planning-what-is-it-and-how-is-it-useful
6. Gartner Research (2019) Optimize utility capital expenditures with asset investment planning solutions. https://www.gartner.com/en/documents/3898772/optimize-utility-capital-expenditures-with-asset-investm
7. Crespo A, Sola A, Moreu P, Gómez JF, de la Fuente A, Guillén A, González-Prida V (2018) Criticality Analysis for improving maintenance, felling and pruning cycles in power lines. IFAC-PapersOnLine. https://doi.org/10.1016/j.ifacol.2018.08.262
8. Crespo Márquez A, Rosique AS, Moreu de León P, Gómez Fernández JF, Diego AG, Fernández EC (2020) Exploiting EAMS, GIS and dispatching systems data for criticality analysis. In: Value based and intelligent asset management. https://doi.org/10.1007/978-3-030-20704-5_7
9. Kolevski G, Gusev M (2010) Analysis of cloud solutions for asset management. In: Gusev M (ed) Conference: ICT innovations 2010, web proceedings, ISSN 1857–7288
10. Hamdaqa M, Tahvildari L (2012) Cloud computing uncovered: a research landscape. Adv Comput.https://doi.org/10.1016/B978-0-12-396535-6.00002-8
11. Metcalfe D, Winter S (2018) Operational risk management software market size and forecast 2018–2038. Verdentix
12. Abualkibash M, Elleithy K (2012) Cloud computing. the future of IT industry. Int J Distrib Parallel Syst 3, N°4, 1–12
13. Maciaszek LA, Skalniak T, Biziel G (2014) Architectural principles for service cloud applications. In: Shishkov B (ed) Business modeling and software design. bmsd 2014. lecture notes in business information processing. Springer. vol 220

Chapter 2
A Review of New Digital Technologies Impacting Maintenance Management

2.1 Introduction

Many companies in our environment are still reading about the value of new technologies for the industry, however many others are already taking advantage of industry 4.0 technologies, reducing their risk in asset management, and responding more quickly to the business demands.

The fact is that many Industry 4.0 technologies are at a point where they really start to make sense for industrial applications. As discussed later, maintenance can be one of the industrial areas most favored by the introduction of new technologies. This opportunity cannot be missed, and it is therefore the responsibility of the maintenance engineers and managers to study the possible implementation of these technologies.

In the following Sections, a set of technologies helping to respond faster to assets maintenance requirements and needs are introduced.

2.2 The 'Internet of Things' (IOT) Technologies

The adoption and development of advanced 'Internet of Things' (IoT) technologies enables important advances impacting on maintenance management. This has been possible thanks to the efforts of open membership organizations such as The Industrial Internet Consortium (IIC), which have promoted the development, implementation and widespread use of interconnected machines and devices and intelligent analytics. Thanks to these type of organizations, many companies are now planning to support the growing IIoT 'Industrial Internet of Things' with cloud servers and an app stores (application components can be found in a cloud environment, integrated with applications managed in private clouds or in data centers) [1].

A. Crespo Márquez, *Digital Maintenance Management*, Springer Series
in Reliability Engineering, https://doi.org/10.1007/978-3-030-97660-6_2

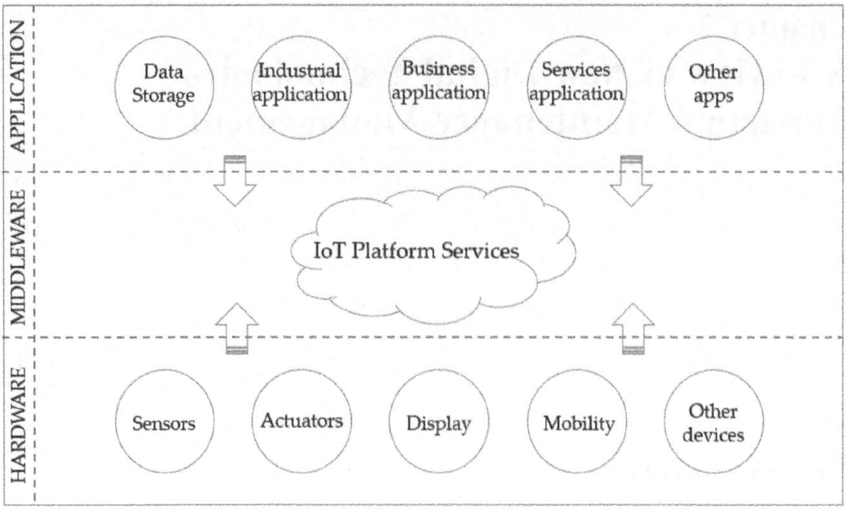

Fig. 2.1 IoT Platform layer

IIoT platforms originated in the form of IoT middleware [2, 3], with the intention to serve as an intermediary between the hardware and application layers. Following this idea, powerful IT vendors such as SAP HANA, GE Predix or RFID4U have shown in their catalogs a structure such as that proposed in Fig. 2.1.

An advantage of IIoT platforms is that it supports integration with almost any connected device and blends with third-party applications used by said device. This independence between hardware and software allows a single platform to manage any type of device in the same direct way.

The new IoT platforms introduce a multitude of interesting features. They provide components for front-end and analytics, on-device data processing, and cloud-based deployment.

The studies by Mineraud et al. [4], Borgia [5] and Perry [6] pointed out that the following platform characteristics should be considered when choosing an IoT platform:

- *Device management*: The platform should store a list of all connected devices with the corresponding operating status. In general, an IoT platform will have to manage from thousands to millions of devices. To operate with low power consumption, light communication protocols must be used.
- *Data collection and storage*: Storage of device data in the cloud is at a higher level in terms of data volume, variety and speed.
- Creation of *virtual machines*: The analysis (in real time, batch, predictive or interactive) of all the data from the sensors connected to the platform provide valuable information.
- *Remote device configuration and control*: Managing the configuration is essential to control the performance of the device.

- *Reliability and data safety*: The IoT platform must ensure reliability in the most demanding environments. The company's most critical functions are at stake and the reliability of the platform must be indisputable.

It is important to highlight the flexibility offered by the best IoT platforms, which allow *adding your own industry-specific components and third-party applications*. This flexibility avoids significant extra costs and delays due to the adaptation of an IoT platform to a specific industry scenario.

2.3 Big Data

In modern APM systems, input data to critical asset management decision-making processes must be collected, evaluated, and analyzed with more than remarkable quality for good results [7]. In an organization that has effective asset management that leads to good decision making, we would expect to find:

- The company identified the data necessary to effectively analyze and manage critical assets, and
- The company specified the quality standards for the data. Data quality can be specified in terms of many attributes according to ISO 9000: 2015, including:
  - *Completeness* (is all of the data to be collected, or only some of it);
  - *Accuracy* (does the data accurately represent reality—particularly a concern when human data input is required);
  - *Timeliness* (is the data available as and when required);
  - *Availability* (is the data readily available to those using it);
  - *Consistency* (are the same definitions and standards applied across the organization).

The time spent by data scientists in the processing and preparation of data represents almost 80% of their work [8] as shown in Fig. 2.2. This work, which includes the collection, cleaning and organization of data is undoubtedly the slowest task faced by data analysts.

Data in modern IoT platforms, when dealing with health monitoring and predictive analytics, is stored and processed in a different way because [10]: its volume is too big, with very high velocity, high variability, low veracity (high noise) and variety is high as well (much diversity). Some examples of analytical functions that big data can perform, thus facilitating data collection and processing are:

- *Complex aggregation analysis*: to process information obtained on different dates or in different locations,
- *Multi-dimensional query and analysis:* to analyze and extract asset data from different viewpoints,
- *Log data analysis*: to control the asset health during its operation,

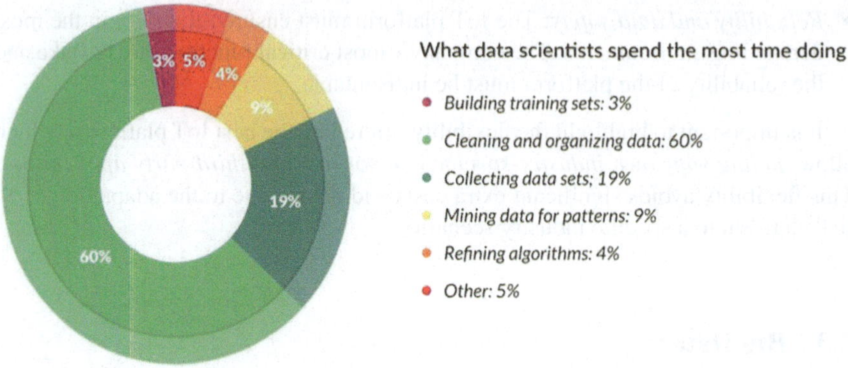

Fig. 2.2 Percentage of time consumed by data scientists [9]

- *Time-window based stream data analysis*: to detect trends in operation and
- *Complex event processing*: to detect failures before they occur.

2.4 Predictive Analytics

Although predictive analytics has been well known for years, the development of new technologies has boosted these techniques. Business CEOs increasingly emphasize the importance of predictive analytics and machine learning as it translates into increased profits.

A survey conducted in 2016 [11] showed that 69% of business managers or data scientists believed that data analysis was essential to the future of their business. Furthermore, 79% of the respondents considered that predictive or prescriptive maintenance was extreme or very important. Well, this trend has been confirmed over the years as shown in Fig. 2.3, even becoming increasingly important in recent years [12].

Depending on the desired results and the available input data, different analyzes can be performed as shown in Fig. 2.4 [1]:

- *Baseline analytics* is used to detect anomalies, and the data used is usually local to the asset itself when it operates under normal operating conditions.
- *Diagnostic analytics* provide very fast results and identify the root cause of the failure, so it requires a prior study of the different states of the failure.
- *Prognostic analytics* is generally used to calculate the remaining useful life of a component, require a multitude of input data and is not as fast as diagnostic analytics.

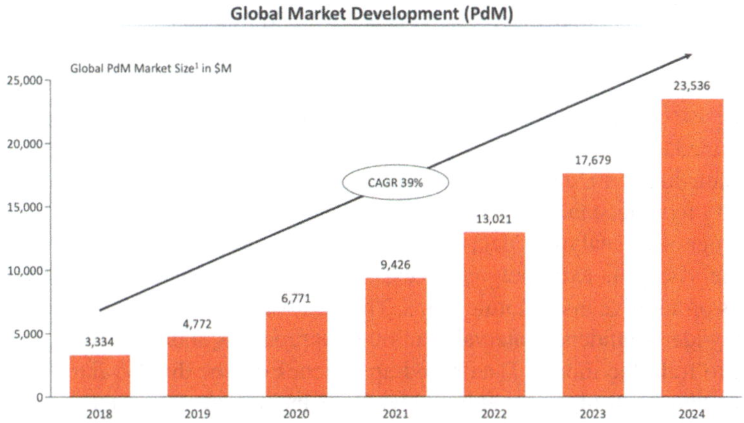

Fig. 2.3 Importance of IA for business managers and data scientists

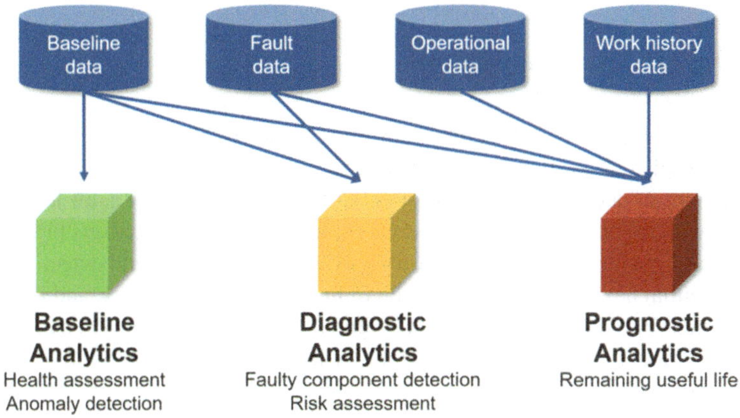

Fig. 2.4 Types of analytics based on their applications [1]

Deployment of analytics generally consists of three steps:

- Training a model for predictive analytics,
- Testing and validating the model with different data and
- Implementing the model in a real scenario.

Important efforts must be developed to reach analytic models providing value by improving asset performance. In most complex engineering assets nowadays, despite the existence of well-developed physical models for a component of a system, the complexity that arises from the combination of elements, and many times the changes in environmental and operational conditions, make it impractical to characterize a complete system through closed mathematical expressions. As a solution, stochastic models are often used to characterize these types of systems.

The application of Artificial Intelligence (AI) and Machine Learning (ML) techniques provide intelligence and flexibility to predictive models and are very valuable for discovering patterns of behavior in the presence of randomness. Both reasons make these techniques increasingly important [13]. In diagnostic analytics, data must be examined to answer the question "Why did it happen?", techniques such as drill-down, data discovery, data mining and correlations are used to take a deeper look at data to try to understand the causes of failures and behaviors.

In prognostic analytics, data-based approaches are gaining ground, especially when the system is extremely complex, and the development of a more accurate model is prohibitive in economic terms. The disadvantages of this approach are that it has a wider confidence interval than other approaches and needs a high amount of data to train the model. Data-based approaches can be divided into fleet-based statistics and sensor-based conditioning. Both include modeling the accumulated damage (or, equivalently, health) and then extrapolating to a threshold of damage (or health) [14], or directly calculating the remaining useful life [15–17].

2.5 Digital Twin Simulations

Digital twins (DT) are virtual replicas of physical devices or processes built to run simulations of real entities that can help to understand, predict, design, etc. This improvement in decision-making throughout the life cycle of an asset leads to the possibility of creating applications based on the DTs of physical assets [18, 19].

By creating faithful twins, the industry will be able to reduce fault detection times, reveal potential hazards, etc., all in search of higher efficiency, safety, and flexibility. In fact, one of the most interesting possibilities is to use simulation to calculate the time that an asset will remain operational. A DT can track its own failures and its own degradation.

By moving from the physical realm to the digital world, companies will be able to try new scenarios and advance potential problems, especially when artificial intelligence is brought into the picture to help decision-making. The faithfulness and predictiveness of digital twins depend on three factors:

- The quality of the models employed. Models need to include all the relevant physics that occur in the real system, but nothing else. Depending on the scale of the problem, these might range from basic linear models of electric circuits, hydraulic piping, vibrations, etc., to complex nonlinear phenomena modelling fluid mechanics, electromagnetics, materials, etc. While more elaborate models can yield accurate predictions, they are more difficult to calibrate and solve in a reasonable amount of time so that their predictions can effectively help to control the real process.
- The quantity and usefulness of the data gathered in the calibration and run-time of the process. In a data-starved world, the internet of things (IoT) is proposing to use ubiquitous sensors to monitor processes and machines, feeding digital twins

with a wealth of information that can be used, first to calibrate them, but especially to go in parallel with modelling and limiting the drift of the latter. For the first time in history, data-driven models can also be integrated with analytical ones, providing black- or grey-box modelling to complement classical analyses.

- The available computational power. As virtual systems grow in size, complexity, and detail, new CPU and GPU parallel networks need to be employed, especially when trying to obtain real-time performance or, even more, exploration of potential and unexpected behaviors.

2.6 Augmented Reality

Augmented Reality (AR) is a form of human–machine interaction that overlays computer generated information (virtual data) on the world environment (real objects) [20]. AR prototypes have demonstrated that this technology can enhance maintenance activities implementation [21], especially when users require additional information [22]. In fact, AR visualization capabilities can transform many maintenance processes. However, AR research in maintenance applications is still at a prototype stage, and a complete integration of AR in maintenance systems has not been pursued yet [23].

New frameworks are now appearing and describing the information to enhance performance and knowledge of maintenance processes through AR capabilities to manage, acquire and display that data properly [24].

2.7 Business Intelligence and Data Visualization Tools

The term business intelligence (BI) refers to a set of technologies and processes providing business metrics for business users [25]. They are frequently needed because many transactional systems do not have a suitable reporting system built in, so producing a given report demanded by the business is difficult.

These BI processes require the creation of a data warehouse with relevant data about transactions taking place and other dimensions that could be of user's interest (named fact tables and other multiple dimension tables), which hold more information about a transaction in question. Put together, all these tables result in a data warehouse. Of course, data warehouse requires creating interfaces from the source data systems or the transactional systems, because when creating some executive reports, one may need to collate reports from multiple systems, to produce something that provides more supporting data than only one system, at a time, can offer.

As the reader may guess, this BI process introduction is not easy and may represent an important effort to solve data-cleansing and harmonization problems, especially in geographically expanded companies, with multiple transactional systems, that could be doing the same thing for different lines of business. At the same time BI processes separate transactional from analytical databases for providing faster calculations and functional data that can embrace multiple transactional applications.

Beside BI processes, data visualization (DV) processes allow transmitting information quickly, effectively, and intuitively. DV turns regular data into information that is commonly provided to the users in three possible ways:

- *Dashboards*: Single page of key metrics in tables, graphs, charts, and numbers, presented for user to easily identify areas requiring immediate attention.
- *Reports*: List of transactions that took place during a certain time, in a specific place, etc.
- *Alerts*: Single metrics or messages that are sent to users. These provide a warning that an event has just taken place exceeding a significant self-defined threshold. Alerts encourage users to action.

Beside existing BI and DV existing opportunities additional data analytics may be required, using data that may or may not exist in the data warehouse and hence may not exist in existing reports. The data analyst might use different tools to dig into the data to produce different views and to identify data patterns. However, these additional tools can also be considered as BI, i.e., BI is not simply a single type of software, but a group of Apps to transform data into information to better monitor and execute their functions.

2.8 Complementary Technologies

Mobility system integration continues to be an important component in advancing asset management strategies and optimizing fieldwork. Most organizations need good mobile capabilities with APM and other complementary technology for offline use. Some organizations are seeking an industry-specific solution such as mobile workforce management (see Gartner's "Market Guide for Mobile Workforce Management Systems for Utilities" [26]). Others opt for a more horizontal mobile solution such as field service management systems (see "Magic Quadrant for Field Service Management" [27]).

Besides data collected through the IoT, mobile technologies can capture inspection data in the field that can easily turn into statistics and condition data. Clearly, these capabilities improve data collection and work efficiency and support workers with access to up-to-date data to know what is happening with an asset—as it relates to the associated location and context—and even why it is happening.

GIS systems can be embedded as a graphical interface, for better planning and routing. For example, some uses of this technology include field mapping for asset maintenance and procurement, considerably improving decision making in the field.

At present, location and context-aware apps are becoming key for productivity improvement. This is not to be confused with GPS-enabled apps on smartphones. GPS provides capabilities such as dynamic rerouting using sourced map tiles (such as Google, Bing or Here). GIS provides capabilities such as viewing layers of infrastructure where the data must be sourced from less "public" sources.

References

1. Industrial Internet Consortium. Industrial internet reference architecture. Technical Report (2015). https://doi.org/10.1371/journal.pone.0158503
2. Schiefelbein UH, Soligo D, Maran V, de Oliveira JPM, Lima JCD, Machado A (2019) Internet of things based on situation-awareness for energy efficiency. iSys: Revista Brasileira de Sistemas de Informação (Brazilian J Inform Syst) 12(1):28–53
3. Xiao Y, Kalyanaraman R, Ylä-Jääski A (2009) Middleware for energy-awareness in mobile devices. Proceedings of the fourth international ICST conference on COMmunication system softWAre and middleware—COMSWARE '09
4. Mineraud J, Mazhelis O, Su X, Tarkoma S (2016) A gap analysis of internet-of-things platforms. Comput Commun 89–90:5–16
5. Borgia E (2014) The internet of things vision: key features, applications and open issues. Comput Commun 54:1–31
6. Perry M (2016) Evaluating and choosing an Iot platform. O'Reilly Media, Inc
7. Hariri RH, Fredericks EM, Bowers KM (2019) Uncertainty in big data analytics: survey, opportunities, and challenges. J Big Data 6:44. https://doi.org/10.1186/s40537-019-0206-3
8. Crespo Márquez A; de la Fuente Carmona A; Antomarioni S (2019) A process to implement an artificial neural network and association rules techniques to improve asset performance and energy efficiency. Energies 12: 3454
9. Edd Wilder-James (2016) Breaking down data silos
10. ISO/IEC JTC 1 information technology. Big data Preliminary Report 2014. Iso/Iec. (2015)
11. Lueth KL, Patsioura C, Williams ZD, Kermani ZZ (2016) Industrial analytics 2016/2017: the current state of data analytics usage in industrial companies. IoT Analyt (2016)
12. IoT-Analytics.com. Predictive Maintenance Report 2019–2024 (2019). https://iot-analytics.com/product/predictive-maintenance-report-2019-2024/
13. Rodríguez F, Fleetwood A, Galarza A, Fontán L (2018) Predicting solar energy generation through artificial neural networks using weather forecasts for microgrid control. Renew Energy.https://doi.org/10.1016/j.renene.2018.03.070
14. Vachtsevanos G, Lewis F, Roemer M, Hess A, Wu B (2007) Intelligent fault diagnosis and prognosis for engineering systems. In: Intelligent fault diagnosis and prognosis for engineering systems. https://doi.org/10.1002/9780470117842
15. Mosallam A, Medjaher K, Zerhouni N (2016) Data-driven prognostic method based on Bayesian approaches for direct remaining useful life prediction. J Intell Manuf.https://doi.org/10.1007/s10845-014-0933-4
16. Mosallam A, Medjaher K, Zerhouni N (2015) Component based data-driven prognostics for complex systems: Methodology and applications. Proceedings of 2015 the 1st international conference on reliability systems engineering, ICRSE 2015. https://doi.org/10.1109/ICRSE.2015.7366504
17. Jia X, Jin C, Buzza M, Di Y, Siegel D, Lee J (2018) A deviation based assessment methodology for multiple machine health patterns classification and fault detection. Mech Syst Signal Process.https://doi.org/10.1016/j.ymssp.2017.06.015
18. Boschert S, Rosen R (2016) Digital twin-the simulation aspect. In: Mechatronic futures: challenges and solutions for mechatronic systems and their designers.https://doi.org/10.1007/978-3-319-32156-1_6
19. Roda I, Macchi M (2018) A framework to embed Asset Management in production companies. Proc Inst Mech Eng Part O J Risk Reliab. https://doi.org/10.1177/1748006X17753501

20. Zhu J, Ong SK, Nee AYC (2013) An authorable context-aware augmented reality system to assist the maintenance technicians. Int J Adv Manuf Technol 66:1699–1714
21. Martínez H, Laukkanen S, Mattila J (2013) A new hybrid approach for augmented reality maintenance in scientific facilities. Int J Adv Robot Syst 10:1–10
22. Webel S, Bockholt U, Engelke T, Gavish N, Olbrich M, Preusche C (2013) An augmented reality training platform for assembly and maintenance skills. Rob Auton Syst 61:398–403
23. Nee AYC, Ong SK, Chryssolouris G, Mourtzis D (2012) Augmented reality applications in design and manufacturing. CIRP Ann Manuf Technol 61:657–679
24. Fernández del Amo I, Erkoyuncu JA, Roy R, Wilding S (2018) Augmented reality in maintenance: an information-centred design framework. Procedia Manufacturing 19:148–155
25. https://www.clicdata.com/blog/what-are-bi-data-visualization-data-analytics/
26. Foust N (2019) Market guide for mobile workforce management systems for utilities. Gartner Doc ID G00355730
27. Robinson J, Rashid N (2020) Gartner magic quadrant for field service management. Gartner Doc ID G00392502

Part II
The Pillars for Digital Maintenance Management Excellence

Chapter 3
Driving the Introduction of Digital Technologies to Enhance the Maintenance Management Process and Framework

3.1 Introduction

The adoption of new digital technologies must be done to facilitate the decision-making processes in the organization's asset management.

These technologies should promote more objective and effective decision-making in asset management, help to recognize risks and uncertainties, focus management on assets, and ultimately, increase the competitiveness of the organization.

However, managers must be very careful. Technology should never take precedence over maintenance processes but should be used to facilitate their implementation and to improve their effectiveness and efficiency in maintenance and asset management.

Managers must design their digital transition in maintenance by looking at the key management areas that can now be better supported, selecting the tools that may be appropriate to improve these areas, and implementing new technologies accordingly.

Without a proper management framework standardization, despite the introduction of digital technologies, managers could misalign with the company objectives, deriving on tailor-made managerial behaviors for each area and conditions. In addition, consensus among these area managers, for instance about budget decisions, could be difficult to reach.

3.2 Starting from the Mm Process and Framework

As mentioned in Chap. 1, managing maintenance properly requires following a process—a course of action—grouping management activities within a series of so-called management building blocks; and then to structure a "framework", grouping techniques that can be used to support decisions-making within each one of these building blocks. In [1] a generic model for maintenance management was proposed

A. Crespo Márquez, *Digital Maintenance Management*, Springer Series in Reliability Engineering, https://doi.org/10.1007/978-3-030-97660-6_3

in Sect. 7.2, for built and in-use assets, and consists of eight sequential management building blocks, as presented in Fig. 3.1.

Each one of the building blocks is, in fact, a key decision area for asset maintenance and life cycle management. Within each of these decision areas techniques, methods and models can be found to order and facilitate the decision-making processes, some of which are presented in Fig. 3.2.

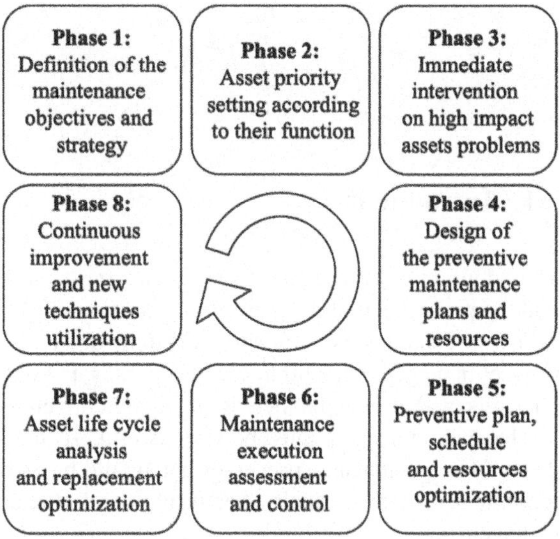

Fig. 3.1 Maintenance management model [1]

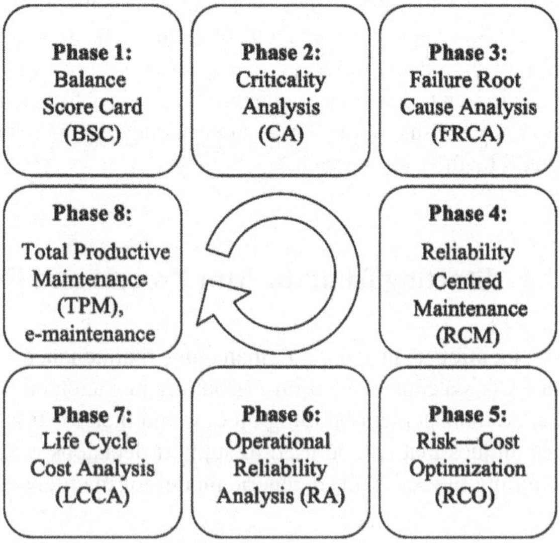

Fig. 3.2 Sample methods and models [1]

Other recent works [2, 3] continue with the development of this framework for maintenance and assets management that has been promoted by the SIM group in the University of Seville over the last few years.

3.3 New Opportunities with Data Availability and Analitycs

Data will now play a pivotal role; additional data, of a different nature, can now be available for asset managers at different stages of the maintenance management process. Considering this data for decision-making within these processes will make a real difference when managing maintenance.

Data such as:

- The aggregate historical data of management, planning, and costs.
- The geographic physical location of the assets, their diverse operational and environmental changes.
- The real-time status of equipment, alarms, thresholds, and degradation patterns.
- The prediction derived from reliability studies and through artificial intelligence.
- Etc.

can now be extracted, prepared, and recorded, for specific decision-making processes, automatically (this is named ETL extraction-transformation-and-load of data).

To that end, new intelligent assets management systems apps (IAMS Apps) are appearing in the market, these apps organize the collection and the analysis of data, for specific decision-making purposes (corresponding to each one of the referred as management building blocks in Figs. 3.1 and 3.2). They work together with ETL apps to generate ad hoc data repositories with a certain data structure model. Apps will ensure that data extracted from selected data sources, is transformed according to information requirements, formats, etc. These data repositories are expected to be placed in databases that are available in the cloud, using IaaS and PaaS Tools.

Then, it is just a question of intelligence, i.e., selecting the specific data for each occasion and purpose, and the proper business rules for each specific decision-making.

IAMS Apps may also interact with additional tools such as simulation tools, providing extra analytical services, and they may add complementary data to the database records with results provided by these software elements.

In addition to this asset knowledge discovering, creation and storing [4], these IAMS Apps are provided by vendors together with business intelligence features or Apps (BI Apps). The BI App is designed for the interaction with the end user and extract database records to present the information according to the reporting needs and end user requirements, on demand or at the time needed by the business. A simple data flow of the process is presented in Fig. 3.3 (adapted from [4]).

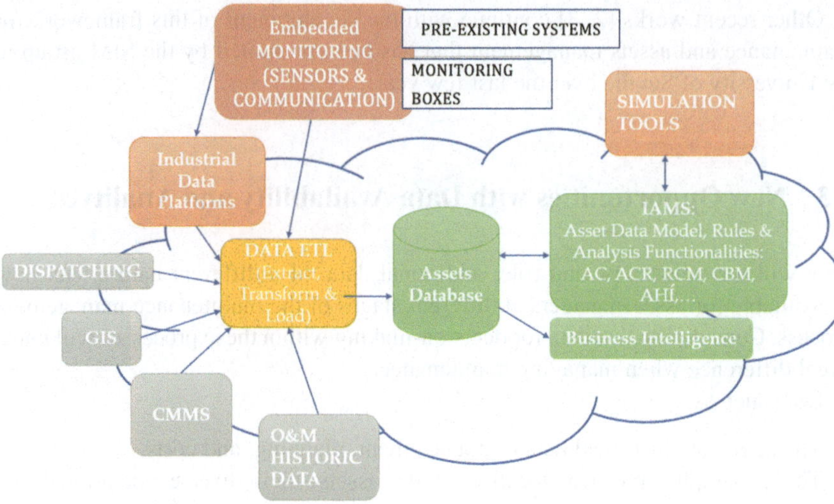

Fig. 3.3 A generic vision of the basic IAMP functional elements [4]

3.4 Visualizing the New Digital Maintenance Management Framework

A graphical tool that eases for the reader the visualization of the overall digital maintenance management framework is a data Input—Process—Output diagram, which can be synthesized and represented as in Fig. 3.4.

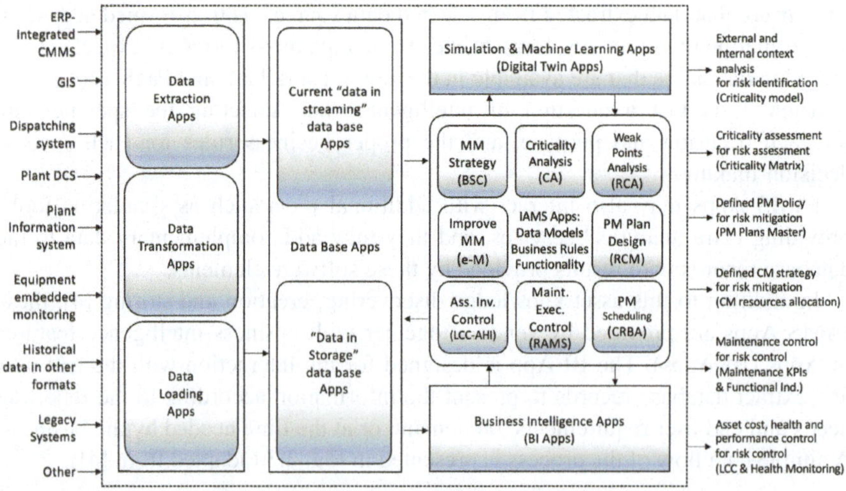

Fig. 3.4 An Input—Process—Output diagram of the DMM framework

In Fig. 3.4, input is raw data from different business systems to be transformed or converted. Systems can be, for instance, ERP systems, dispatching systems, GIS systems, DCS, etc. Wherever relevant asset information is stored in the business [5, 6].

The process is then divided into different building blocks (like ISO 14224:2016 approach [7]), each of the blocks representing a system: ETL systems, Database Systems, IAM Systems, AI systems and BI systems. Each one of these systems will have a certain function (or group of functions) to ensure an effective and efficient digital maintenance management.

The different systems will handle a very precise and predetermined data model to generate an output. These outputs can be for different purposes: to identify risk in assets, to assess that risk, to mitigate the risk, etc. Whatever is needed to control risk for the business in assets normal operations. Each one of these outputs is related and assigned to a different IAM App.

DT Apps and BI Apps are very important supporting Apps within this framework. They may interact with the IAM Apps to allow the introduction of powerful data analytics and visualization tools.

Figure 3.4 is a simple schematic of a fairly complex digital maintenance management system.

3.5 The Need for a Business Asset Data Model

As the reader may guess, the data model becomes critical to each one of the processes that the different Apps support. We can certainly talk about a suitable asset data model for a given business, once we have identified those apps required to support the relevant processes for asset management in that business, and we take into consideration their data requirements.

Boundaries shall avoid overlapping between different Apps datasets, but some overlapping may be difficult to avoid, however, such case(s) shall be identified and treated appropriately during the data analyses.

The next Chapter will study a process to structure data models for intelligent asset management apps. A framework is presented to understand the different sources of data required and their interrelation.

References

1. Crespo Márquez A (2007) The maintenance management framework. Models and methods for complex systems maintenance. Springer, UK, ISBN 978-1-84628-821-0. https://doi.org/10.1007/978-1-84628-821-0
2. Crespo Márquez A, González-Prida Díaz V., Gómez Fernández JF (eds) (2018) Advance maintenance modelling for asset management. Techniques and methods for complex industrial systems. Springer International Publishing. ISBN 978-3-319-58044-9

3. González-Prida Díaz V, Parra Márquez C, Crespo Márquez A (eds) (2021) Cases on optimizing the asset management process. Techniques and methods for complex industrial systems. IGI Global. ISBN 978-1-79987943-5
4. Crespo Marquez A, Fernandez JFG, Fernández PMG, Lopez AG (2020) Maintenance management through intelligent asset management platforms (IAMP). Emerging factors, key impact areas and data models. Energies 13. https://doi.org/10.3390/en13153762
5. Crespo Márquez A, Rosique AS, Moreu de León P, Gómez Fernández JF, Diego AG, Fernández EC (2020) Exploiting EAMS, GIS and dispatching systems data for criticality analysis. In: Value based and intelligent asset management. https://doi.org/10.1007/978-3-030-20704-5_7
6. Crespo A, Sola A, Moreu P, Gómez JF, de la Fuente A, Guillén A, González-Prida V (2018) Criticality Analysis for improving maintenance, felling and pruning cycles in power lines. IFAC-PapersOnLine. https://doi.org/10.1016/j.ifacol.2018.08.262
7. ISO 14224:2016. Petroleum, petrochemical and natural gas industries—collection and exchange of reliability and maintenance data for equipment

Chapter 4
The Definition of the Asset Data Model

4.1 Introduction

To ensure the functionality of the Apps, they must structure the asset data model and configure the databases where all relevant data for the corresponding analysis will be stored. To this end, in this Chapter the framework introduced by Guillen et al. [1], for CBM purposes, is generalized.

The referred framework classifies the different nature of the source data required for the AM decision-making process, grouping the data in 5 processing blocks. In that precise paper, that dealt with with the data structure for CBM solutions, data elements were grouped as presented in Fig. 4.1.

The proposed framework comprises five blocks. Four of them correspond with maintenance management processes blocks (blocks 1, 2, 4 and 5) and one block of information resources for the processes (block 3). The sequence of blocks can also be interpreted as the basic process that the user will follow to analyze any CBM solution.

This Chapter claims that this framework can be easily generalized by changing the terms used in the definition of the last two blocks of the framework, as presented in Fig. 4.1. Where block 4, comprises the analysis done for a certain asset management process and block 5, the final AM process decision-making (as in Fig. 4.2).

Through this blocks sequence, and its conceptual representation in UML (Unified Modeling Language) class diagram, any IAMS App can be developed with different functionalities, following their concepts and relations as standard, and facilitating the computational implementation in a sustainable way.

The key to this data model is to link the physical structure elements and functional logic elements and identify parameters required for the selected analysis, to provide an accurate support to the considered decision-making process.

The pursued results of the framework implementation are: to perform a better problem analysis, collect, and incorporate expert knowledge, improve, and automate the decision-making and increase the value of decisions as continuous improvement.

A. Crespo Márquez, *Digital Maintenance Management*, Springer Series
in Reliability Engineering, https://doi.org/10.1007/978-3-030-97660-6_4

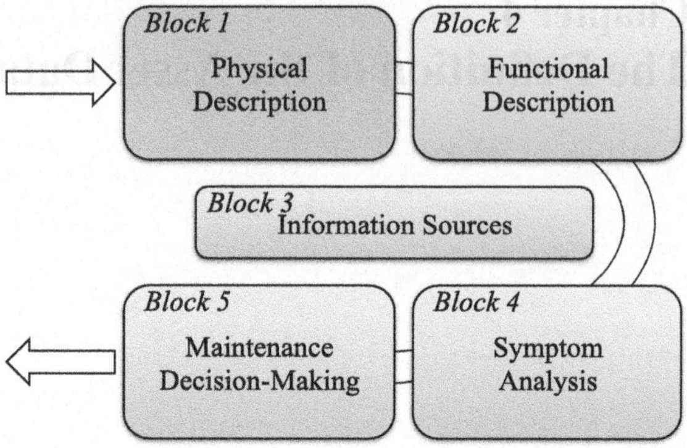

Fig. 4.1 Processing blocks in the proposed framework [1]

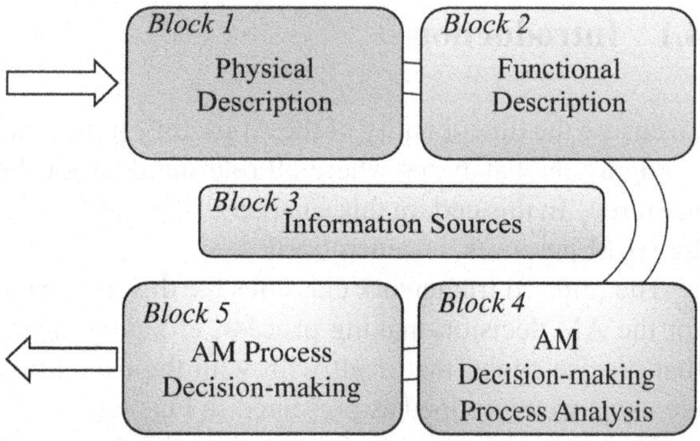

Fig. 4.2 Generalization of the framework proposed by Guillen et al. [1]

4.2 Methodology Proof of Concept: Data Model for a Criticality Analysis App

Criticality analysis (CA) is understood as the methodology that allows establishing a hierarchy of the assets of a company, creating a structure that facilitates making successful and effective decisions. Specifically, in the energy industry, this analysis takes on special importance. Asset management in the energy sector is focused on assets, therefore it is key to have all the assets of a company hierarchized. This hierarchy allows companies to withdraw resources from less important assets and put maximum resources into the most critical assets, which can cause serious problems for the company.

A process to rank the hierarchy of electrical network assets according to criticality was presented by Crespo et al. [2]. To this end, a Criticality Analysis App could be offered to an end customer. The App would support the CA process which is designed to be applied to a massive number of assets (over 200.000 power lines in our example), and that would need to be re-elaborated every six months, for a proper asset criticality tracking. The question is: How can a data model, for this business process purpose, be elaborated?

Proceeding according to the previous Section and to the original work [1], a detailed representation of the data model proposed for this criticality analysis, using a UML class diagram, is presented in Fig. 4.3.

In a UML diagram, classes of entities are in rectangles, linked to include information related to each other. This link includes an expression on the triangle that describes the specific role that entity class plays in an association (aggregation, composition and reflexive associations), and at both ends a number indicating the number of instances of a class related to one instance of the other class (that is, 0..*—zero to any number) [3]. This schema allows a comprehensive view of the data structure, elements and blocks, and the relationships among them. The reader can appreciate the data requirements and the way the relationships among data elements are structured.

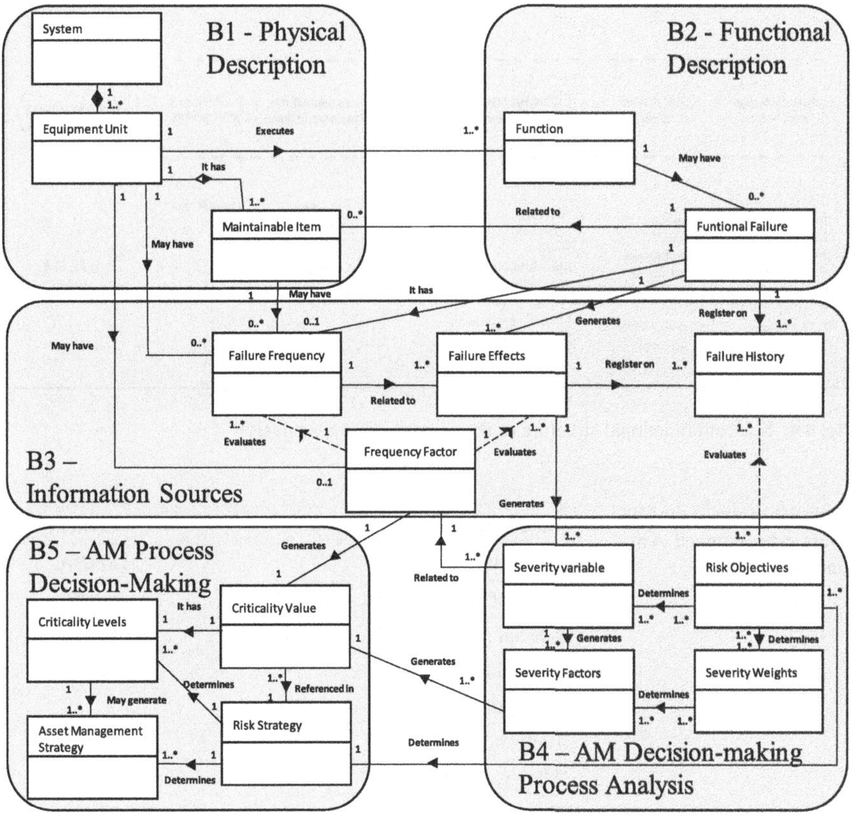

Fig. 4.3 Data model supporting the assets CA decision-making process

Figure 4.3 illustrates the relationship among data that is required for the criticality analysis, through this figure the link to the experimental process and data can be illustrated:

- It shows in Block 1 how a system is composed by one or more than one equipment units, that will be placed in specified functional locations and properly geo referenced (data in this block), each one executing one or more functions.

  For the power lines example, 6 groups of functional locations were considered as maintainable items (see Fig. 4.4), and geo referencing these locations was included in a set of data attributes, as presented in Table 4.1.

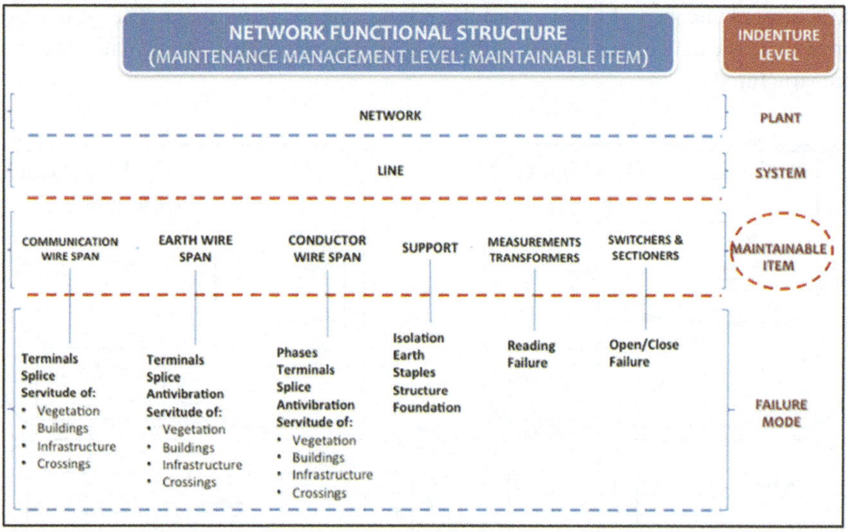

Fig. 4.4 Selected functional structure of the network for the analysis

Table 4.1 Data in the GIS layers to be assigned to a functional location

Data layers in the GIS	Content
Fire risk zone	Yes/No
Place of public interest	Yes/No
Special protection zone (animals)	Yes/No
Natural Park	Yes/No
Vegetation fraction covered (%)	%
Railway crossing	Yes/No
Main road crossing	Yes/No
Populated zone	Yes/No
High frequency of persons area	Yes/No
Other network crossing AT, MT, BT, Phone	Yes/No

Table 4.2 Frequency factors used in the power lines example

Annual frequency failure	Classification	Management definition	Frequency factor
$2 \leq f$	Very high	Inadmissible failures	2
$1 \leq f < 2$	High	Recurrent failures	1,5
$0{,}5 \leq f < 1$	Medium	Acceptable failures	1,2
<0,5	Low	Possible failures	1

- In Block 2 for Functional Description, each function may have different identified functional failures. Each function may have one or more functional failures and these failures will be caused by one or more maintainable items.
- Block 3 depicts how when the functional failure occurs, the description of the effects of failure, history and frequency are recorded in a specific activity management system, and that data can be used later in Block 4.

  For the power lines example, the CMMS had recorded the failures of each functional location, and the criticality App just assigns a factor of failure frequency to each functional location according (see Table 4.2 with the factors used in the example):

For the same example power lines' failure effects were classified according to their impact on industrial safety, environment, quality of service, availability of the network and maintenance cost (see Table 4.3). The criteria selected to be considered for the analysis were defined according to the objectives of the company. The most important in that case was the quality of the service (35%), followed by industrial security (30%). Logically, for each company, specific factors and weights will be considered according to their own objectives.

- In this example, the Block 4 AM decision-making analysis, refers only to the asset's criticality analysis. For this kind of analysis, the contribution of each functional failure must be measured in terms of risk, that requires capturing the frequency of the failure on one hand (deriving in a Frequency Factor) and estimating the Severity of the failure on the other hand (deriving in a Severity Factor). Frequency times severity results in a measure of failure risk that must fit within the company risk objectives. Both parts must be matched for each specific functional failure.

In the example of the power lines, the company's CMMS is connected to the GIS (geographic information system) and assets locations are recorded when assets are registered as shown in Table 4.1. Data for Frequency Factors for each Maintainable Item and each Functional Failure that are collected from CMMS. At the same time, the "fault location" functionality can be used by a dispatch center to provide information

Table 4.3 Failure severity criteria and levels used in the power lines example

Industrial safety (30%)		Environmental (12%)		Quality service (35%)		Availability (14%)		Maintenance costs (5%)	
Catastrophic: very serious effect with damage to the network or fatalities	**100**	**Catastrophic: non-recoverable effect**	**100**	**Catastrophic: loss of continuity in electric service for more than 20,000 customers**	**100**	Very high: total loss of capacity of the line	20	Very high: maint. Cost > 30.000 €	5
High: high impact to the net or serious injury or severe damage to third parties (>300,000 M€)	30	High: recoverable impact on vulnerable area	12	High: the same for 2,000–20,000 customers impacted	35	High: total loss of capacity of the line in grid or branch network with redundancy	10	High: 30.000 € > Maint. Cost > 6.000 €	4
Medium: average impact, or damage to third parties (>50,000€)	18	Medium: outside impact in non-vulnerable area	6	Medium: the same for 200–2,000 customers impacted	12	Medium: partial loss of line capacity with possibility of remote maneuvering	5	Medium: 6.000 € > Maint. Cost > 1.000 €	3
Slight	1	Slight	1	Slight	1	L Slight	1	Slight	1

Table 4.4 Sample environment criteria effects conversion (using GIS data)

ORIGINAL		AUTOMATED	
ENVIRONMENT: 12% CATASTRIFIC:CA HIGH: A MEDIUM: M LOW: B		**ENVIRONMENT: 12%** CATASTRIFIC:CA HIGH: A MEDIUM: M LOW: B	
NON-RECOVERABLE EFFECT ON VULNERABLE AREA (PROTECTED AREAS)	**CA 100**	**RISK OF FIRES IN AREAS ENVIRONMENTALLY QUALIFIED AS SCI, SPAB OR NATURAL SPACES, THAT HAVE COVERAGE FRACTION >=70%**	**CA 100**
RECOVERABLE EFFECT ON VULNERABLE AREA	A 12	RISK OF FIRES IN AREAS ENVIRONMENTALLY QUALIFIED AS SCI, SPAB OR NATURAL SPACES	A 12
OUTSIDE IMPACT IN NON-VULNERABLE AREA	M 6	COVERAGE FRACTION <>0	M 6
SLIGHT IMPACT OR NO IMPACT	B 1	SLIGHT IMPACT OR NO IMPACT	B 1

about the potential number of affected clients when a failure occurs at a specific point in the network. Failure severity data (factors, their weights, and scores of the failures for effects on these factors) is normally placed in companies' custom-made Apps for that purpose [5]. Thanks to these possibilities, the assessment of the potential impact of a functional location failure can be done using automated rules according to the functional location GIS data. For instance, see in Table 4.4 how a functional location failure impact on "Environment" can be derived automatically from GIS data.

– In Block 5, data for all failures frequency and severity for a given equipment unit is transformed into a criticality value and then classified according to risks level (company assets criticality levels: high, mid or low criticality). This makes the definition of actions easier according to an existing Risk Strategy to fulfil the company Risk Goals. In consequence, in Block 5, the decision-making process for each asset (equipment unit) is to determine the correspondent strategy to be applied to the asset to fulfill the company risk strategy, and also to make clear what the priority is for the execution of these asset strategies in a systematic way.

Table 4.5 Criticality limits

Criticality	Value range
Critical	91–200
Semi-critical	50–90
Non-critical	1–49

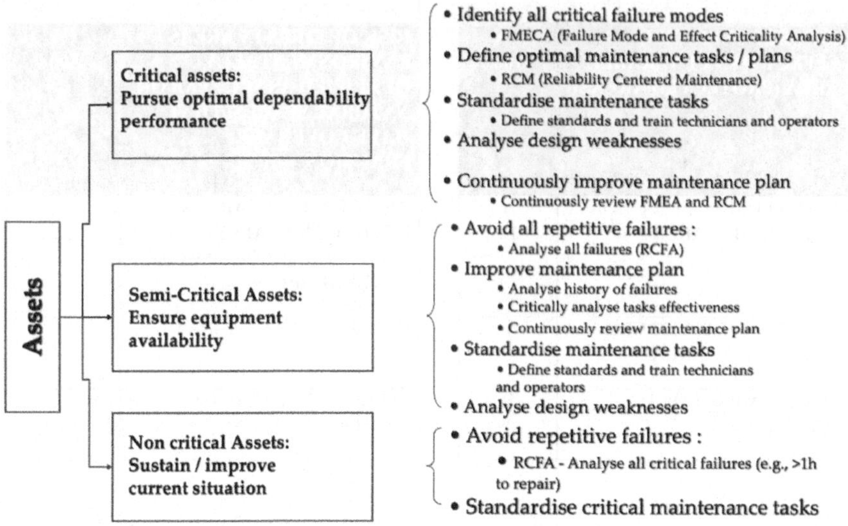

Fig. 4.5 Assets' strategy, according to their criticality, to fulfil company's goals

For instance, in the powerlines case the determination of the criticality limits was established by the business as in Table 4.5.

Immediately after this decision and the automatic evaluation of the asset's criticality, the actions to accomplish the company goals can be established as in Fig. 4.5, as a starting point for another App utilization.

Notice how Blocks 1 and 2, as physical and functional inventory of the assets, will be common data blocks for all the Asset Management Apps in an IAMP. While data in Blocks 4 and 5 are clearly specified and modelled for each type of Asset Management Decision-Making App.

When the information mentioned above is available, a set of simple rules to automate the criticality analysis can be used. Failure consequences, for all the criteria previously selected and for each asset, can be assessed immediately.

For a complete description of the process followed in the case study and for specific elements of the network, the reader is referred to Crespo et al. [4].

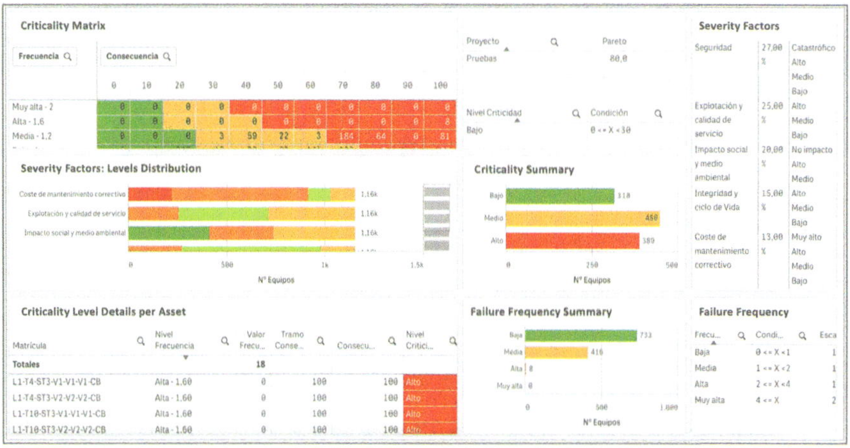

Fig. 4.6 Results (only headings in english) of the BI in the asset criticality analysis APP

Using a business intelligent App, we can present information about the criticality analysis process to interact with the end user and according to the reporting needs and end user requirements, on demand or at the time needed by the business (see an example in Fig. 4.6).

References

1. Guillén AJ, Crespo A, Gómez JF, Sanz MD (2016) A framework for effective management of condition based maintenance programs in the context of industrial development of E-maintenance strategies. Comput Ind 82. https://doi.org/10.1016/j.compind.2016.07.003
2. Crespo A, Sola A, Moreu P, Gómez JF, de la Fuente A, Guillén A, González-Prida V (2018) Criticality Analysis for improving maintenance, felling and pruning cycles in power lines. 2018. IFAC-PapersOnLine. https://doi.org/10.1016/j.ifacol.2018.08.262
3. Ciccozzi F, Malavolta I, Selic B (2019) Execution of UML models: a systematic review of research and practice. Softw Syst Model 18:2313–2360. https://doi.org/10.1007/s10270-018-0675
4. Crespo Márquez A, Rosique AS, Moreu de León P, Gómez Fernández JF, Diego AG, Fernández EC (2020) Exploiting EAMS, GIS and dispatching systems data for criticality analysis. In: Value based and intelligent asset management. https://doi.org/10.1007/978-3-030-20704-5_7
5. Jia X, Jin C, Buzza M, Di Y, Siegel D, Lee J (2018) A deviation based assessment methodology for multiplemachine health patterns classification and fault detection. Mech Syst Signal Process 99:244–261

Part III
Developments in Advanced Maintenance Management Apps

Chapter 5
Advanced Asset Performance Management (APM) and Asset Investment Planning (AIP) Systems

5.1 The APM Toolbox Approach to Asset Maintenance

In his Market Guide for Asset Performance Management Software, Gartner Research describes the process that asset-intensive industries are following concerning asset management. This management discipline evolves from simple maintenance to a business operations core competency. Gartner argues [1] that APM software is at the core of this change and that CIOs must understand new observed market direction and support these practical asset management strategies.

APM products are delivering many valid strategies available for improving reliability and managing maintenance. These should not be thought of as "good" and "bad," but rather as different approaches to the same overall issue, targeting different circumstances [1]. The objective of APM within a maintenance strategy roadmap is to have a broad array of skills and tools and then to match the needs of different asset classes to those tools in a toolbox approach. Just as a maintenance worker would not use a hammer for every job and would have an array of tools available for any given job, an organization will need to leverage many different skills and tools for asset maintenance.

In the toolbox approach to asset maintenance, organizations should map out the best strategies for the different assets within their businesses. Operations will have needs that fall into different categories within this structure, and the vendors that may be suggested won't fulfill all possible levels and variations. Therefore, IT departments, in conjunction with the business users, should determine a viable roadmap to expand the available systems from where they are now to a full multilevel capability. It is important to note that most APM vendors do not offer all levels of APM maintenance strategies. Thus, some organizations may need more than one APM product, depending on the complexity of their businesses, the types of assets and their asset maintenance goals.

APM applications can provide different functionalities and are usually included in different categories or in different modules of the software packages. Each of these Apps requires data that can come from different data stores. Table 5.1 includes

A. Crespo Márquez, *Digital Maintenance Management*, Springer Series in Reliability Engineering, https://doi.org/10.1007/978-3-030-97660-6_5

Table 5.1 A definition of APM capabilities and categories given in [1]

Capability	Processes and Tools
Asset strategy and risk management	• Data collection and aggregation from EAM systems. Integration with major EAM systems for updating maintenance plans. Various analysis techniques and tools for calculating risk and assessing criticality, including: • Weibull analysis—statistical distribution of asset life data from a representative set of sample units to predict the life of an asset • Risk-based inspection (RBI)—analysis methodology and process that requires qualitative or quantitative assessment of the probability of failure (PoF) and the consequence of failure (CoF) associated with each equipment item • Fault tree analysis (FTA)—a deductive failure analysis method that models the pathways within a system that can lead to failures or undesired results • Mechanical integrity—management of critical process equipment to ensure it is designed and installed correctly, and that it operates and is maintained properly (that is, all elements are fit for service) • Safety integrity level (SIL) analysis—a method to indicate the tolerable failure rate of a particular safety function • Library of failure modes and recommended practices
Reliability-Centered Maintenance (RCM)	• Data collection and aggregation from EAM systems • Root cause failure analysis—actions taken to determine why a particular failure or issue exists and correcting those causes • Failure mode and effects analysis (FMEA)—a method to identify where and how an asset might fail and to assess the relative impact of different failures
Predictive maintenance	• Statistical modeling/regression analysis of physically observable characteristics in a piece of equipment, recorded over a time period • Neural network analysis—computational data model that can capture and represent complex input/output relationships • Machine learning to determine the related characteristics that are observable in a failure of equipment or parts so as to use that as the basis of future failure indicators • Proprietary algorithms • Monte Carlo simulation
Condition-based Maintenance (CbM)	• Rule engines—to determine and record tolerance levels for physically observable characteristics in a piece of equipment

Source Gartner (June 2019)

the APM Apps capabilities and their processes and tools according to Gartner [1] to illustrate the aforementioned. This classification could be discussed, for instance, the RCM capability could well be considered a process or tool to elaborate an adequate strategy for maintaining the asset and managing its risk.

In this Chapter, a review of some OEM software divisions' and IT companies' products is done. It is an exercise that is carried out with the purpose of finding those Apps that can cover what are considered basic asset management processes to improve asset performance. In the same way, some aspects that have to do with the implementation of these platforms are analyzed.

The information included in the Chapter is obtained from the information on the software, provided by its suppliers, in addition to the experience of the real-life implementation of these platforms.

5.1.1 GE Digital's Predix Asset Performance Management

Predix Platform was developed by GE Digital [2]. In December 2018, GE announced a new company by merging GE Digital with GE Power Digital Solutions and GE Grid Solutions.

GE saw a great market opportunity and created the Predix platform, which can connect physical asset data to powerful analytical tools and industrial applications, to improve the performance of assets, reduce costs and develop new services.

Predix APM is comprised of Predix Platform-powered SaaS and on-premises Apps with the intention to "increase the reliability and availability of assets while optimizing maintenance costs, minimizing risks, and reducing total cost of ownership (TCO)". The suite works in all OEMs and all industries and covers the full range of industrial asset needs: APM Health, APM Reliability, APM Integrity and APM Strategy.

In Table 5.2, first column, the basic AM key company business processes are presented. The heading of the next four columns of the table, are the names of the Predix platform modules (APM Health, APM Reliability, APM Integrity and APM Strategy). Within those modules in the current platform structure, the Apps that are relevant to support the key AM business processes are identified. The table is therefore a cross reference table linking AM key processes for the company, to the corresponding Predix platform App. This is done considering a user that wants to have a basic tool kit supporting a maintenance model such as the one depicted in Chap. 3.

5.1.2 SAP APM Solution

A well-known global IT vendor offering APM tools is SAP. In the second quarter of 2019 the 1905 version was released, consisting of four modules:

(1) Predictive Maintenance and Service.
(2) Asset Strategy and Performance Management.
(3) Predictive Engineering Insights; and
(4) Asset Intelligence Network.

Table 5.2 Correlation of business AM basic processes with predix platform apps (Taken from [3])

Sample basic AM process (application)	Apps big modules in GE predix			
	APM health	APM reliability	APM strategy	APM integrity
Criticality analysis for assets hierarchy	Asset criticality analysis: uses default indenture level and severity factors. The user can select factors			
Root cause analysis		Root cause analysis: based on the PROACT® RCA methodology		
PM and inspection plan design			Reliability centered maintenance: uses SAE standard JA1012, "evaluation criteria for RCM processes"	Risk based inspection. using the API 580/1 methods to derive risk and optimized inspection plan
PM and inspection strategy review/management			Asset strategy management: allows to create a strategy to optimize equipment performance to achieve business goals	Inspection management: to manage inspection plans, document the condition of an asset on a mobile device, and track inspection recommendations
Predictive/condition-based asset management		Digital twin blueprints and Predictive diagnostics: leverages predictive diagnostics analyzing data and detects and diagnoses well in advance		
Asset performance analysis		Reliability analysis: collection of tools that apply reliability engineering principles		

(continued)

Table 5.2 (continued)

Sample basic AM process (application)	Apps big modules in GE predix			
	APM health	APM reliability	APM strategy	APM integrity
Life cycle cost analysis			Life cycle cost analysis: It provides the ability to balance costs, plan budgets, collect performance data and build contingency plans	
Spares analysis		Reliability analysis/spares analysis: Monte Carlo simulation is used to spares analysis		
Health index	Asset health manager: a graphical summary of an asset's health based on information stored in various records is presented			

All these modules are integrated with SAP's EAM portfolio and are becoming popular in a wide range of industries with locations mainly in Europe and North America, although with a presence in most regions globally.

These modules include, for instance, RCM in SAPs Asset Strategy and Performance Management product. Collaborative APM as commercial services in SAP Asset Intelligence Network, or capabilities for deploying custom algorithms and user defined analytics in SAP Predictive Maintenance and Service, or vibration analysis and simulation as a service, as well as embedded simulation models from partner ANSYS is included in SAP Predictive Engineering Insights. Moreover, SAP offers offline mobile use through SAP Asset Manager.

SAP offers integration to its own IoT platform and other third-party data sources (for instance one of the most popular is OSIsoft's PI). Data collected via sensors can be loaded and stored in SAP HANA, the company's in-memory DBMS. SAP's CBM capability can be provided through the SAP PM module (EAM solution) or through the APM solution for Predictive Maintenance and Service. In this new solution SAP offers predictive analytics and machine learning, as well as Predictive Engineering Insights for simulation models.

Product interfaces are established for EAMs such as SAP ERP Central Component (ECC) and SAP S/4HANA, GIS like Esri, while no interface is provided for any AIP. SAP APM solutions System Integrators are well-known companies, for example, Accenture, Deloitte, etc.

5.1.3 Siemens APM Products

Siemens APM products have a strong presence in Europe, North America, and the Middle East, with clients from a great deal of industrial sectors. Siemens offers four different APM products, all of them interfacing with their MindSphere IoT Digitalization platform, across several divisions:

(1) MindSphere (latest release is version 3, released in January 2018).
(2) APM for Power Plants version 3.0, released in December 2018.
(3) Power Diagnostics Center (PDC); and
(4) Condition Monitoring and Diagnostics.

MindSphere offers CBM capabilities through its Visualizer App, predictive analysis through its Advanced Analytics App, RCM in its Product Intelligence App and RBI in the RBI App. MindSphere is deployed primarily as a hosted solution, which included releases on AWS in January 2018, Microsoft Azure in May 2018, and Alibaba in September 2018. There is also an on-premises version available through a partnership with ATOS that was released in April 2018.

APM for Power Plants, an APM solution for power generation, is part of the Siemens' MindSphere digital platform, and combines Bentley's asset performance management expertise with Siemens's operations and maintenance experience, fleet data and domain knowledge. Predictive analytics are built into asset models and delivered as starter content.

The PDC product is a very utility-specific App that focuses on abnormal operating conditions detection of power and operation equipment to improve plant operations reliability and availability.

Siemens provides an offline mode in all the products, that are aligned with ISO 55000 and ISO 27001 industry standards.

Bentley has partnered with Siemens to provide a combined product called PlantSight (launched early 2019) for process industries. PlantSight leverages Siemens' MindSphere, COMOS and XHQ with Bentley's iTwin Services, OpenPlant, AssetWise and Interoperability Services to provide plant visualization and analysis capabilities.

Interfaces are established for EAMs such as SAP and Oracle, are not reported for GIS nor for any AIP system. SAP APM solutions System Integrators are well-known companies like: Atos, Bentley, Deloitte, TCS, Software AG, IT Vizion, etc.

5.1.4 *IBM APM Portfolio*

The current IBM Maximo APM portfolio was released In October 2018. At present, IBM has two products offering APM capabilities:

(1) Maximo Enterprise Asset Management, positioned in Watson IoT group, and
(2) Maximo Asset Performance Management. The branding of the Maximo APM solutions was updated in March 2019, and it combines:

- Maximo APM—Health (formerly Asset Health Insights),
- Maximo APM—Predict (formerly Predictive Maintenance Insights), and
- Maximo APM—Assist (formerly Equipment Maintenance Assistant).

The Maximo Asset Management products offer an offline mode of use and can be found in many asset-intensive sectors for which Maximo APM has released multiple modules during the last few years.

Maximo APM provides basic condition monitoring, asset criticality analysis, health assessment and FMECA tools. Some of these products have also enhanced industry-specific packages, with a focus on the energy and utilities sector. In fact, Maximo APM for Energy & Utilities (version 3.0 released in January 2019) provides a set of predictive models for utilities equipment and predictive and AI-assisted maintenance capabilities. Concerning RCM, IBM is partnering with MaxGrip and GE Digital. Also, in 2016, IBM acquired the rights to resell Fluke Accelix Connect2 Assets product to connect APM solutions to industrial automation systems.

IBM Maximo APM suite works seamlessly with Maximo EAM and other EAM offerings, as an interface with Esri GIS and not with AIP systems as of present. This IBM suite integrators are IBM Services, Deloitte, Capgemini and Accenture.

5.2 The Asset Investment Planning (AIP) Solutions

AIP solutions aide rethinking asset strategies over replacing assets. A wide variety of organizations are seeing more complex revenue streams and are increasingly under pressure to do more with less, forcing them to rethink asset utilization. AIP systems take data from EAM and APM systems—primarily on asset's health, criticality, and depreciation—and analyze the data to determine spending alternatives. AIP helps organizations make objective, data-driven decisions to optimize cost and minimize operational risk over the full investment life cycle of an asset, from predicting asset conditions to funding for repair, reallocation, or replacement (see [1]).

The timeliness of each option of intervention on the asset, or its own substitution, is determined through an economic analysis, and according to certain strategic considerations defined by the user, such as budgetary and regulatory restrictions (for example, ISO 55000). Indeed, every organization needs a mechanism by which they determine the value of an investment decision if they intend to optimize the use of their scarce resources.

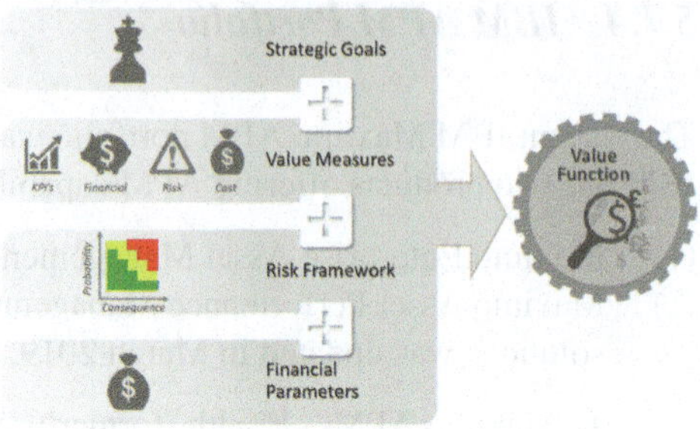

Fig. 5.1 Value function for the systematic evaluation of investment proposals [4]

AIP solutions try to provide a well-defined value framework to enable a value-based decision-making aligning with the organizational and strategic objectives of the organization to consistently quantify investment value across the business [4].

The quantification of the value of very different investment proposals is difficult and can become a daunting task. AIP solutions develop a value function that can help streamline this process. The strategic goals or objectives of the organization will be translated into several value measures that can be quantified and weighted. These measures, together with the risk framework and the financial parameters (i.e., inflation rate, discount rate) are used to develop a function that fairly reflects how the organization will value the benefits claimed by each proposed investment (Fig. 5.1).

In these current times, investments in maintenance and capital for fixed assets are more critical than ever. Executives of companies with capital-intensive assets must make many decisions that can significantly affect the income and expenses of their companies, both in the medium and long term.

The current economic downturn leaves little room for error in determining asset investment strategies and evaluating their associated risks. It is therefore important to have AIP-type Apps to support these processes and refine, as much as possible, decision-making. Software providers such as Copperleaf or Cosmo Tech are impacting the market with solutions that are every day less specific to regulated industries, and that expand quickly to adjacent asset-intensive industrial sectors [5].

In the next sections a variety of these solutions are reviewed. The intention is not to be precise in their elements content, but to offer a general overview of the tools they use for optimization and their integration with other AM systems possibilities.

5.2.1 Copperleaf Decision Analytics

Headquartered in Vancouver, Canada, Copperleaf® delivers software solutions to companies managing critical infrastructure (electricity, natural gas, water, government, oil and gas, transportation, and other asset-based industries) enhancing their decisions analytics.

The Copperleaf Suite combines asset data with financial information to enhance long-term planning, budgeting, and forecasting decisions, enabling organizations to maximize the value of investment portfolios and drive strategic objectives.

The solution contains a set of different modelling capabilities that can be divided in two parts: the Asset Management part and the Investments part.

- Asset management models comprises a whole series of advanced asset modeling capabilities such as:
  - Ability to define any type of degradation curve (Condition vs Time and Probability of Failure vs Time), e.g., distributions Weibull with size and shape parameters.
  - Calculation of Condition / Probability of Failure / Health Score / Health Index at the component level and automatic aggregation for the asset set.
  - Functionality to model dependency and impact on the degradation of different assets close to each other.
  - Etc.
- Investment performance models: Includes capabilities such as:
  - Investment performance prediction: Capability to forecast the most probable real execution of the Investment Portfolio and its deviation from the Original Plan, as well as the uncertainty bands based on the historical data of execution of projects/investments using Monte Carlo simulations.
  - Etc.

Currently Copperleaf® solutions are based on advanced data analytics tools and optimization algorithms (MILP type linear optimization: Mixed Integer Linear Programming) providing market leading performance. This applies to both the Investments part and the Asset Management part. For example, the Predictive Analytics module can provide the optimal sequence of interventions on any asset considering its complete life cycle.

In recent times, Copperleaf has made a decisive commitment to increase the capabilities of AI/ML throughout the entire Suite and now constitutes one of the main lines of work of the product roadmap with some new modules already available.

Copperleaf Asset does not normally act as an asset master but imports the structure and technical information of assets from the EAM, however, in many cases, it does act as a master of attributes or characteristics related to Condition/Risk modeling (e.g., curves of degradation and useful life for each type of asset, values of environmental consequence/security/reliability of supply, unit costs for each type of intervention on each type of asset, etc.).

Copperleaf software can be implemented on-premises in the client's IT infrastructure (use of application servers/standard database servers without special requirements), in an internal client cloud (on Amazon Web Services, Azure, etc.) or on Copperleaf's own cloud (on AWS).

5.2.2 The Cosmo Tech Hybrid Approach

Cosmo Tech is a spin-off of ENS Lyon, founded by Michel Morvan and Hugues de Bantel in 2010. It is considered a world reference in modeling, simulation, and optimization of industrial systems. Their product strategy is based on the construction of digital twins, which makes them different from other competitors in the concept of product and offer to the market. The idea of the firm is to support, through digital twins, decision-making processes at a strategic level. It offers the ability to virtually test customer action plans before they are implemented. The models are intended to help obtain an optimal and executable strategy and to be able to define an operational plan with full transparency. Its main clients are in the manufacturing, automotive, energy and transportation sectors.

Cosmo Tech's proprietary Complex Systems Modeling Language (CoSML) is core to model and simulate physical assets, industrial systems, procedures, and processes. CoSML represents a real system with reference to four key components of that system: entities, hierarchies, interactions, and dynamics.

As part of its Cosmo Tech Asset Solution, the company has developed an AIP App that allows the creation of a Digital Twin of a specific infrastructure: it is about obtaining a dynamic replica of all operating systems, including physical assets, limitations, processes and strategies, human resources and regulatory limits. From that moment on, it is possible to carry out simulations to evaluate different asset investments and replace options during the selected planning period. The application allows the manipulation of key aspects such as budgets, configuration, criticality of assets, degradation models, maintenance frequencies, changes in the technical location, laws, etc. It is clearly an App designed to manage the complexity of business asset portfolios, which can present results for business risk obtained at different indenture levels. A tool that pursues to transmits more confidence in decisions about optimal asset renewal and investment plans in the short, medium, and long term.

Cosmo Tech solutions explore the combination of simulation and ML from their inception. Their simulation digital twin solutions make this hybridization operational for the most complex industrial problems. The possibility to train reduced ML models from simulation results, or to explain through simulations those correlations detected by ML models, results in a very powerful hybrid modelling approach, that can exploit the strength of each digital technology. In fact, this hybrid approach emerges as a growing trend in the scientific community. Researchers at MIT, OpenAI and the Fraunhofer Center for Machine Learning in Germany are just some of the groups already exploring this path [6].

References

1. Foust N, Steenstrup K (2019) Market guide for asset performance management software, IDG00388410. Gartner Research
2. GE Digital (2018) Predix asset performance management. https://www.ge.com/digital/sites/default/files/download_assets/Predix-asset-performance-management-brochure.pdf
3. Crespo Marquez A, Fernandez JFG, Fernández PMG, Lopez AG (2020) Maintenance management through intelligent asset management platforms (IAMP). Emerging factors, key impact areas and data models. Energies 13. https://doi.org/10.3390/en13153762
4. IAM (2016) SSG 6 & 7 Capital investment Ops & maintenance decision making
5. Voulgaropoulos V (2021) Asset investment planning: what is it and how is it useful? Verdantix, London. https://www.verdantix.com/blog/asset-investment-planning-what-is-it-and-how-is-it-useful
6. Gros P-A, Hospital N (2021) ML and simulation. Unlocking the power of a hybrid approach. http://Cosmotech.com

Part IV
Technical Challenges in the Use of Digital Technologies for Maintenance

Chapter 6
The Non-ergodicity of Assets

6.1 Introduction to the Problem

Different assets, by definition, have characteristics of non-ergodicity [1, 2]. What does this mean? A system is ergodic if for a controlled experiment, studying a given process in several experiments running in parallel, correspond to studying the same process over time in a single experiment [2] (see Fig. 6.1). In other words, the mean of the sample of experiments converges to the expected value over time of a single experiment. In industrial asset fleets these properties are not fulfilled. If a machine component fails several times, the mean properties of these failures do not necessarily converge to the mean properties of all the failures of that component in the fleet. This means that asset fleets are often non-ergodic [3].

In the context of this book and Chapter, when considering the possibility of using artificial intelligent tools for assets maintenance, this can seriously impact the possibilities for collaborative learning, i.e., the ability of assets to use information from other assets to update their own failure detection and predictions.

Inter-asset differences may cause a predictive model, relevant to a particular asset, to be non-applicable to another asset in the same fleet. For example, in a fleet of bearings in a train, the year of production, bearing model, operation and preventive maintenance history will influence the properties and behavior of a bearing in each train position. This illustrates an effect encountered frequently when performing anomalies detection, failure modes classification and prognostics in an asset fleet: lack of ergodicity [3].

Thus, for example, for the case study that we present later in this chapter, it is necessary to find out whether the bearings in the fleet are similar enough so that the knowledge learned from one bearing in one position can readily transfer to another one in a similar position. As mentioned, this cannot be true due to existing variations in make, operational history, location, and maintenance history. Therefore, a strategy that considers using data coming from other assets to perform prediction for a given asset must be proven.

A. Crespo Márquez, *Digital Maintenance Management*, Springer Series
in Reliability Engineering, https://doi.org/10.1007/978-3-030-97660-6_6

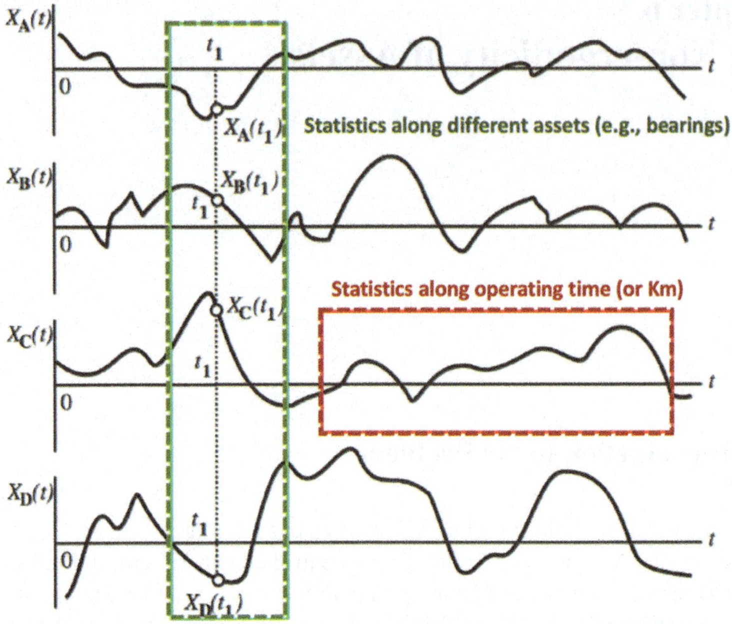

Fig. 6.1 The non-ergodicity of assets may avoid collaborative learning (Adapted from [4])

The lack of ergodicity is an inevitable problem in real scenarios that is usually combated by using very large data sets in which each possible example is represented, and in which a machine learning algorithm can automatically discern between each different case. Sometimes, when such data is rarely available (because all efforts are put into reducing the number of failures), more ad hoc approaches must be followed [3, 5].

6.2 A Case Study for Train Bearings

This analysis focuses on axle bearings of a passenger trains fleet operating on Kazakhstan's railways. To get an idea, Kazakhstan is the ninth largest country on Earth yet has a population of just 16 million people. Kazakhstan's railway network spans over 15,000 km and uses the 1520 mm Russian track gauge, a legacy of its Soviet past. It has a strongly continental climate, with cold winters almost everywhere, and summers from moderately warm to steamy. The enormous distance of the oceans causes the Asian continent to become very hot in the summer and cool somewhat in the winter, so that the records of heat and cold are extreme: in the north they can reach −50 °C in winter as much as +40 °C in summer, while in the south the temperature can range from −35 to +45° C, reaching in the worst case an annual thermal oscillation close to 90°.

Each train axle, in the train model of this case study, is equipped with four axle bearings, two inner and two outers. The temperature of each bearing is controlled by a PT-100 type temperature probe. Each PT-100 probe conforms to the requirements of standard EN 60,751:2008 [6] and allows continuous monitoring of bearing temperature by the train control monitoring systems (TCMS). The TCMS will perform data capture of bearing temperatures with sampling frequency every minute, with the associated variables of date, time, outdoor temperature and train speed. The system allows 30,000 samples to be stored (corresponding to 500 operating hours of the train) and then downloaded for further analysis. In compliance with the applicable regulations EN 50,155: 2018 (Railway applications—Rolling stock—Electronic equipment), the correct operation of the control and supervision system under these conditions is guaranteed, without disturbing the operation of other on-board systems. In parallel, the train electronic control unit (ECU) will receive the signals on-line and process them according to certain control/safety rules (that will be described later), to send notices to the train supervision post, to control speed limit or to open the loop for a train immediate stop.

The fleet of trains under consideration has a total of 16 trains, the same model of OEM train, but grouped in 5 different train configurations (i.e., with more or less cars of a given car type) and two bearings models from different OEMs (BOEM1 and BOEM2), which are used indistinctly at any bearing functional location of any train. Also, axle bearings normally have several major maintenance activities at the workshop, restoring their functionality, and they reach a life span around 4 to 5 million km. After each one of these interventions, they can be placed at a different functional location of the train, in a different train of the fleet. Furthermore, each axle bearing family may require different minor/greasing maintenance activities until major maintenance at 800.000 km.

An ANN algorithm was selected because of the best ratio of correlation coefficient vs total time (training time + scoring time), assuming a minimum required correlation factor of 0.9600 [1]. Moreover, as a complementary result of this analysis, critical factors for prediction were identified. For instance: when predicting $T1$, the most important attributes to consider in the predictive model are $T2$, $T3$, $T4$ and Ta. A similar result was obtained for the prediction of the rest of the bearing positions, relevant factors for prediction were found to be the temperatures of the others axle bearing plus the outside temperature Ta.

At this time the analyst must deal with the characteristics of non-ergodicity of the bearings fleet. It is necessary to find out whether the bearings in the fleet are similar enough so that the knowledge learned from one bearing in one position can readily transfer to another one in a similar position. A strategy that considers using data coming from other assets to perform prediction for a given asset must be proven.

Clearly, it can be expected that the strategy that uses data from other assets to train an asset specific prediction model, may reduce the prediction capability of the model. However, if minimum requirements are fulfilled, the number of prediction models for bearings would be significantly reduced, and that would easy the management of the subsequent CBM plan.

In this Section, the ANN model will be trained and computed with different input data sets, with data coming from:

- the bearing itself, located in a position of a given train axle,
- all the bearings in the same axle position, regardless the axle of a train, or
- all the bearings in the same axle position, regardless the axle of the train and the train of the fleet.

The result of this "model training analysis" will be three different models, named ML Models 1, 2 and 3, respectively:

- ML Model 1: Data set from a Ti bearing of an axle j of a train k
- ML Model 2: Data set from bearings Ti of any axle j of a train k
- ML Model 3: Data set from bearings Ti of any axle j of any train

With $i = 1, 2, 3, 4$ bearing positions; $j = 1, 2, \ldots, 21$ axles; and $K = 1, \ldots, 16$ trains.

In the next Subsections, once the three ANN models are built trained with the above-mentioned data sets, their precision can be tested and compared to explore the possibility of overcoming the non-ergodicity of the bearing and train fleets.

6.2.1 *Axel Bearing Position Model (ML Model 1)*

Creating a single axle building temperature prediction model requires the selection of data for specific bearings over time (Data set from a Ti bearing of an axle j of a train k). In this case study the data received (number of files) for the different trains and years is presented in Table 6.1.

To deal with this large amount of data, the first data preprocessing phase consisted of extracting a sample of 10% of the initial population of the data per each year, maintaining the correlation between variables so that it remains representative. The databases generated had an average of 30,000 data records per train and year. This was done using the *Loop Files*, *Append* and *Store* rapid miner operators as presented in Fig. 6.2. The idea was to use 2018 data to refine models training and then use 2019 data to search for specific bearing problems where to test performance of the models built.

Three specific bearing were selected for the case study: two in train TRA (axle AA T1 and Axle AB T2) and one in train TRB (Axle AC T4), the reason for that was to select a minimum of three bearings with failure history, recorded at the CMMS, that could be later used to have information about the suitability of the prediction models fit. Results of the ANN models for prediction with 2018 data (70% random for training and 30% random for testing) are presented in Table 6.2.

Initially, a warning was placed on $T2$ ML Model 1 performance, with very low correlation compared with previous and expected results. As explained later, that

Table 6.1 Number of data files received per train

Date	TR1	TR2	TR3	TR4	TR5	TR6	TR7	TR8	TR9	TR10	TR11	TR12	TR13	TR14	TR15	TR16
2018	150	101	100	111	58	51	135	146	158	108	163	46	82	89	156	102
2019	58	53	34	49	36	41	70	72	86	70	85	39	42	37	85	28

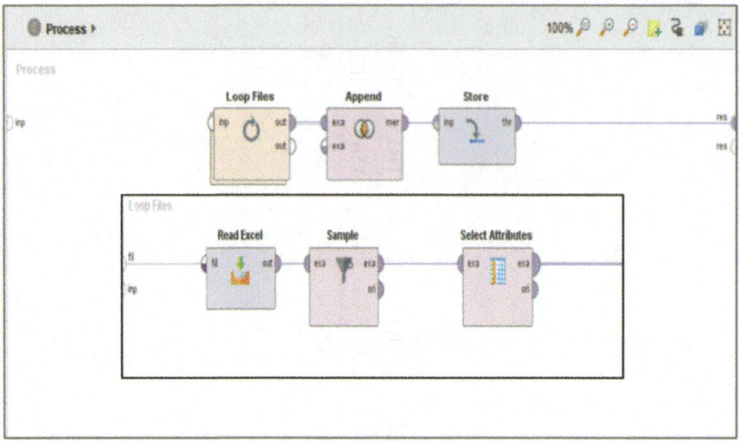

Fig. 6.2 Data aggregation process 1, per Train and Axle, with RapidMiner operators

Table 6.2 ANN prediction models' results per bearing, axle and train analyzed

Year 2018 data	RMSE	Absolute error		Relative error		Correlation coefficient	Avg. Prediction	
		μ	±σ	μ	±σ		μ	±σ
Bearing T1, Axle AA, TRA	3.438	2.29	2.564	19.25%	38.67%	0.980	20.663	16.817
Bearing T2, Axle AB, TRA	5.041	3.60	3.599	28.44%	76.74%	0.595	24.741	16.568
Bearing T4, Axle AC, TRB	3.524	2.78	2.249	20.80%	51.85%	0.974	27.544	25.510

was a consequence of abnormal bearing temperature behavior for a very long period over that year (almost 30% of the time) before the bearing in that axle position and train was replaced. The performance of the M1 model algorithm was found to be very good except for $T2$.

6.2.2 *Train Bearing Position Model (ML Model 2)*

Building this model requires the aggregation of training data coming from any axle of a train, and for a specific bearings position, over time (the data set will have data from bearings Ti of any axle j of a train k). To do so, a second phase of data preprocessing is required, involving standardization of variable names, eliminating any reference to axle in data. New databases were generated having an average of 800,000 records

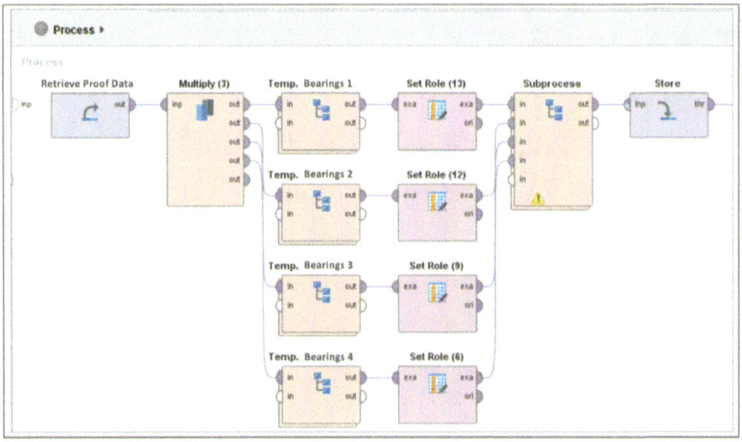

Fig. 6.3 Data aggregation process 2, per Train, with RapidMiner operators

Table 6.3 ANN prediction models' results per bearing and train analyzed

Year 2018 data	RMSE	Absolute error		Relative error		Correlation coefficient	Avg. Prediction	
		μ	$\pm\sigma$	μ	$\pm\sigma$		μ	$\pm\sigma$
Bearing T1, TRA	3.345	2.34	2.387	22.24%	47.61%	0.978	19.535	16.093
Bearing T2, TRA	4.121	2.53	2.251	22.75%	52.56%	0.969	19.696	16.565
Bearing T4, TRB	3.089	1.89	2.443	15.64%	35.30%	0.982	20.119	16.161

per train and year. This was done using the *Retrieve, Multiply, Rename variable Subprocess, Set Role, Union Subprocess* and *Store* rapid miner operators as presented in Fig. 6.3. Results of the ANN models for bearing temperature prediction per bearing and train with 2018 data (70% random for training and 30% random for testing) are presented in Table 6.3.

The performance of the M2 model algorithm was found to be above expectations (>0.96 correlation coefficient) for the three bearing positions.

6.2.3 Fleet Bearing Position Model (ML Model 3)

The third phase of data pre-processing consists of joining the 16 databases for 2018 and the 16 for 2019, generating 2 databases for training and production (containing data sets from bearings Ti of any axle j of any train). The 2018 database has a total of 14,000,000 records and 2019 with 6,800,000 records. This was done using the

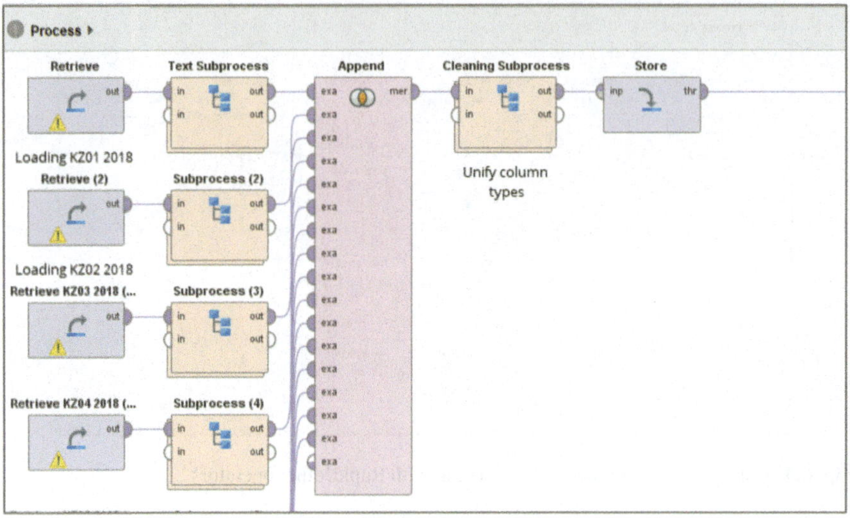

Fig. 6.4 Data aggregation process 3, per fleet, with RapidMiner operators

Retrieve, Text Subprocess, appened, Set Role, Cleaning Subprocess and *Store* rapid miner operators as presented in Fig. 6.4.

Results of the ANN models for bearing temperature prediction per bearing position and fleet with 2018 data (70% random for training and 30% random for testing) are presented in Table 6.4.

Again, the performance of the M3 model algorithm, universal for all trains, was found to be above expectations (>0.98 correlation coefficient) for all four bearing positions, meeting the business *Smart Maintenance Engineering and New Projects* department requirements.

Table 6.4 ANN prediction models' results per bearing and fleet

Year 2018 data	RMSE	Absolute error		Relative error		Correlation coefficient	Avg. Prediction	
		μ	±σ	μ	±σ		μ	±σ
Bearing T1	3.306	2.03	2.258	22.33%	48.01%	0.982	16.891	15.829
Bearing T2	3.193	1.93	2.541	20.60%	46.14%	0.980	16.625	15.982
Bearing T3	3.105	1.87	2.481	19.75%	43.54%	0.981	16.850	15.992
Bearing T4	2.917	2.30	2.122	22.59%	49.48%	0.983	17.163	15.854

6.2.4 Testing and Validating the Models with Different Data

This very interesting test was undertaken to compare to the generic fleet bearing position model (ML Model 3) prediction, to the single axle bearing model (ML Model 1) one. The fleet model was used to test its performance using only specific single bearing data, for the same 2018 dataset. Results are registered in Table 6.5.

The results obtained are rather interesting:

1. As expected, prioritizing the learning of similar machines over self-learning leads to greater error (or average difference between collective error and self-learning error). If the mean difference in error is positive, in theory self-learning would be the best strategy. In these cases, the differences of average errors are (We compare the results of Table 6.5 with those for AE in Table 6.2):

- $\Delta(AET1_{Similars} - AE\_T1_{Self}) = 2.60–2.29 = 0.31$ °C
- $\Delta(AET2_{Similars} - AE\_T2_{Self}) = 4.14–3.60 = 0.54$ °C
- $\Delta(AET4_{Similars} - AE\_T4_{Self}) = 2.86–2.78 = 0.08$ °C

In practice, the error of collective learning algorithm ($AETi_{Similars}$, obtained with ML model 3), compared to that of self- ($AETi_{Self}$, obtained ML with model 1), is always less than 1 °C. That means that the possibility of reducing the number of models for the entire fleet to four, one per bearing, is acceptable according to OEM requirements, and simplifies the CBM solution of the train rolling system.

2. Surprisingly, ML Model 3, for $T2$, outperforms ML Model 1. This can be explained by the possibility to train the ML Model 3 with a data set which contains a higher number of records with normal bearing behavior. Furthermore, it has been verified that the training of model 1 has been carried out in a period in which it is not behaving as expected, coinciding with a maintenance operation in 2018. A better training improves prediction results for periods when the bearing is operating as expected.

Table 6.5 ANN bearing fleet model applied to a single bearing location prediction

Year 2018 data	RMSE	Absolute error		Relative error		Correlation coefficient	Avg. Prediction	
		μ	±σ	μ	±σ		M	±σ
Bearing T1, Axle AA, TRA	4.091	2.60	3.094	24.91%	32.29%	0.972	18.223	15.750
Bearing T2, Axle AB, TRA	5.88	4.14	4.180	30.12%	64.59%	0.954	24.673	16.560
Bearing T4, Axle AC, TRB	3.877	2.86	2.021	33.85%	68.27%	0.970	17.970	15.633

6.2.5 *Conclusions of the Case Study*

Thanks to the big data processes developed, it could be practically demonstrated that the performance of a universal model per bearing position (ML 3, ANN predictive model,) was above expectations (>0.98 correlation coefficient) for all four bearing positions. This made it possible to overcome the non-ergodicity of the assets possible and permitted the development of four prediction algorithms for the entire train bearings CBM plan only.

References

1. Márquez AC, de la Fuente Carmona A, Marcos JA, Navarro J (2020) Designing CBM plans, based on predictive analytics and big data tools, for train wheel bearings. Comput Ind 122:103292. ISSN 0166–3615. https://doi.org/10.1016/j.compind.2020.103292
2. Pathria RK, Beale PD (2011) Statistical mechanics—computer simulations. In: Statistical mechanics, 3rd edn. Elsevier, Academic Press, pp 637–652. https://doi.org/10.1016/B978-0-12-382188-1.00016-5
3. Palau AS (2020) Distributed collaborative prognostics. PhD dissertation, University of Cambridge
4. https://www.quora.com/How-can-I-detect-whether-my-data-are-ergodic-and-stationary
5. Dhada M, Jain AK, Herrera M, Hernandez MP, Parlikad AK (2020) Secure and communications-efficient collaborative prognosis. IET Collab Intell Manuf
6. EN 60751:2008 Industrial platinum resistance thermometers and platinum temperature Sensors

Chapter 7
The Curse of Dimensionality

7.1 Introduction to the Problem

The curse of dimensionality is a phenomenon that appears in Machine Learning models when algorithms must learn from an ample feature volume with abundant values within each one [1], see Fig. 7.1. Reaching samples with each combination of values when training would be very complicated. Thus, it can happen (as it will be appreciated later in this book) that classifier or regressor accuracy first improves including more dimensions but then could even decrease.

This is a common problem when selecting the correct subset of features for the design of a CBM strategy. A good choice may result in models working faster and producing more accurate predictions.

For instance, to design CBM strategies for a compressor maintenance, algorithm can be based on DCS and CMMS data. In these cases, the identification of the necessary variables per failure mode, to detect the incipient failures (point P in the PF curve used in CBM), and then the definition of rules for real time failure mode prediction when approaching the functional failure (point F), are needed. However, screening among the large number of features available to model the condition of a compressor becomes a serious problem. It may require technical knowledge in data mining and artificial intelligent techniques [2] in addition to the relevant failure physics.

This Chapter deals mainly with this problem and to that end several feature selection and feature selection ranking (FSR) methods are considered. These methods are basically algorithms which include wrappers and filters, and they can provide a ranking of all the analyzed features. A very important issue of these methods is to realize the unconstrained feature selection of the Machine Learning algorithm to be later applied.

A. Crespo Márquez, *Digital Maintenance Management*, Springer Series in Reliability Engineering, https://doi.org/10.1007/978-3-030-97660-6_7

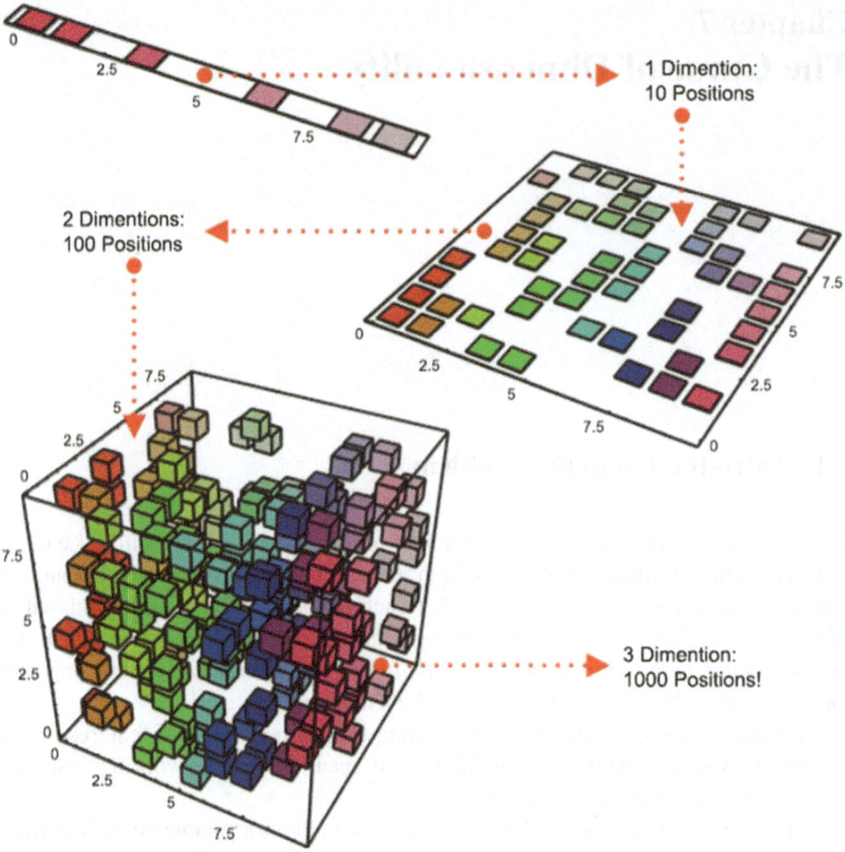

Fig. 7.1 A vision of the curse of dimensionality problem (taken from [18])

7.2 Feature Selection Methods

The data mining process (DM) includes the following steps:

(i) Data pre-processing, where data must be clean, selected and transformed previously to algorithm application.
(ii) Data mining, in which the intelligent algorithm application is realized searching the knowledge in form of patterns.
(iii) Pattern evaluation, the evaluation of patterns or models is done through different metrics of interest that identify those truly interesting patterns, validate some hypothesis or new data with a certain degree of certainty.
(iv) Knowledge representation, where the produced knowledge is displayed in a friendly manner.

Within the pre-processing step, feature selection also known as variable selection, is the process of selecting a subset of relevant features to use in model construction.

Feature selection methods can be divided into the following three broad categories:

- Filter. These methods apply a feature selection algorithm that it is independent from the subsequent learning procedures performed on the dataset.
- Wrapper. These are techniques using an objective function (classifier or regressor) based on an optimization criterion to select features, and
- Embedded methods. Techniques applying a filter method before applying a wrapper method [3].

Other techniques such as Principal Component Analysis (PCA) reduce the number of features with a drawback in which the original features are combined into principal components. However, these principal components from a practical application, could imply difficulties in interpretation if the combinations are complex.

From the point of view of the generated output, the feature selection methods can be grouped into two categories:

- Subset of features. This method selects a minimum set of features that satisfies an evaluation criterion, and
- Ranking of features. These ranking methods are widely used in data mining (DM) for three reasons:

  a. They are used in a pre-processing stage, thus reducing the dimensions of the feature space used to build the model.
  b. Their high computational scalability that requires only computing sorting a set of scores; and
  c. Their statistical scalability [3].

Testing and evaluation of features, with respect to a class variable, can be done using, for instance, the following FSR techniques:

- *Chi Square* (FSR-CS) [4] is based on the calculation of the χ^2 score for each feature.
- *Shannon Entropy* (FSR-SE) [5] determines data appearance likelihood through histograms over feature distribution values and their occurrence frequencies.
- *Differential Shannon Entropy* (FSR-DSE), based on previous algorithm but considering variance in feature distributions according to the range of values.
- *Euclidean Distance-Based Entropy* (FSR-EDSE), through this variable the features are ranked [6] by a clustering tendency based on the distance among different instances.
- *Fisher* (FSR-Fisher) [7] scores different features by the square of the difference between the mean of negative and positive instances divided by counting the squares of the standard deviation of the positive and negative instances. Through this the score of positive and negative correlations have equal importance.
- *F-Score* (FSR-FScore) [8], as the Fisher method, discerns in the binary classification sets considering positive and negative samples with the same weight.
- *Gain Ratio* (FSR-GR) [9], ranks features also discerning the instances classification by the information gain provided by each feature.

- *Information Gain* (FSR-IG) [10]: this FSR method evaluates each featured independently, calculating the information gain to determine their importance and relevance to each class.
- *Mutual Information* (FSR-MI) [11] weighs features according to the mutual information of each input j in instances classification.
- *ReliefF* (FSR-ReliefF) [12] weighs features depending on patterns similarity. ReliefF is considered not only efficient but also successful as a feature estimator through its robustness against noisy data.
- *Recursive feature elimination* (FSR-RFE) [13] weighs features by an outside estimator, removing the less importance from all the set recursively continuing the elimination process with each iteration.
- *t-Test based ranking* (FSR-Ttest) [9]: this FSR method weights are based on the t-test, using mean and variance of the samples results failure and no-failure.
- *Unbalanced correlation score* (FSR-CUB) [7] scores by calculating the correlation coefficient between the features.

7.3 The Use of Classifiers to Identify Failure Modes

Once the feature selection process is done, classifiers are used to discriminate failure modes. Classifiers are ML models identifying objects based on a certain number of features (which may be previously ranked with an FSR method or not). Many contributions can be found proposing the use of different classification algorithms for fault/failure detection problem. A review on the use of classifiers for fault diagnosis of electric equipment can be found in Contreras-Valdes et al. [19].

As an example of classifiers, the following methods can be used to identify failure modes:

- K-nearest neighbors (k-NN) [3]: supervised machine learning algorithm used for classification (discrete value as an output, e.g., "rod drop failure mode" or "not rod drop failure mode"). By K-NN, every data point is classified as the most common class among its k nearest neighbors. It can be assigned weights to the contribution of each neighbor depending on their distance to the data point to be classified. Depending on the problem, the distance between two data points is calculated with a certain method. The number of neighbors "k" must be adjusted as too few (not enough neighbors to avoid odd data points or the noise) or too many neighbors (there are so many data points that the algorithm considers data points which are too far to be considered similar and the boundaries between classes become less distinct) makes the algorithm fail. When $k = 1$ this algorithm is called the 1-nearest neighbor. Its main advantages are that it is one of the easier-to-be-implemented algorithms, it is simple, it does not require to build a model and tune it. The main disadvantage is that the computer processing dependence according to the number of features increases.
- Naive Bayes (NB) [14]: probabilistic machine learning algorithm used for classification using the Naive Bayes principle. There are three types of NB classifiers:

multinomial (frequency of appearance of discrete features), Bernoulli (Boolean features) and Gaussian (continuous variables). In this research the Gaussian NB has been applied. NB is mostly used with qualitative variables although it can be used with quantitative ones. In contrast to the previous, this is easy to implement and highly scalable. However, it requires independent variables.
- Random forest (RF) [14]: ensemble machine learning method used for classification and regression among other applications. RF assumes that various models uncorrelated, the trees, would make a better decision that any one of them separately. Therefore, the key factor for the model to work properly is the low correlation between models, especially regarding the errors committed by every tree. RF consists of several individual decision trees, each one of them predicts the class of a data point and, finally, the most voted class becomes the model prediction. To create independent decision trees, the algorithm uses bagging and random selection of features. Every tree is trained separately with a random sample with replacement and therefore is independent from the others since they are sensitive to the data they are trained on. Additionally, trees use different subsets of features as well, which makes them even more independent.

As mentioned above, classifier selection for this problem is not the goal of this Chapter but feature selection. However, a certain number of classifiers are considered here to see implications of feature selections on a given ML classifier model, and to prove that implications of proper feature selection will be important regardless the ML model selected.

7.4 Case Study: Cryogenic Reciprocating Compressors

7.4.1 Introduction to the Industrial Process

Liquefied natural gas (LNG) is industrially stored at −160 °C and between 100 and 200 mbarg approximately. At these conditions, LNG is at its boiling point and therefore a significant quantity of it evaporates. Regasification terminals and LNG tankers must deal with this BOG generation and regulate their tanks pressure to reduce and eliminate, if possible, the amount of it burned in their flares as it is a valuable substance. Its composition basically consists of methane (>97%) and nitrogen (~2%) depending on the LNG composition. During transportation, BOG may be used as fuel, but onshore plants try to re-condensate it or send it out straight. During the LNG storage, BOG generation depends on many factors: pressure and composition of the LNG, the storage tank itself (age, type, shape, construction materials, capacity, level of LNG stored…) and even weather conditions as tanks are designed to work with a few hundred mbarg overpressure. In addition, there are on and off operations that generates significant amounts of BOG such as LNG tankers loading/unloading operations, LNG trucks loading operations and re-circulations to keep lines and equipment cool to have them ready for their use. Therefore, the quantity of BOG to deal with varies noticeably.

There are two main equipment to recover the BOG, cryogenic reciprocating compressors (from now on referred as cryogenic reciprocating compressors or simply as compressors) and re-condensers. The latter is usually a single equipment but the former normally have back-up ones. Regular use compressors are sized to deal with the day-to-day BOG generation rate with a safety margin. Back-up ones helps to deal with the BOG generation or recovery necessities peaks. This way compressors may be smaller and consequently less expensive.

The cryogenic reciprocating compressors studied in this case study are electrical and have two stages, working with a compression ratio between 5,6 and 9,9 approximately, depending on process conditions. The average temperature gap between suction and delivery is 160 ºC roughly. A general drawing is shown in Fig. 7.2. They are fed by BOG collectors which connect the whole facility, and its pressure depends mainly on tanks pressures. BOG is vaporized at −160 ºC but from the point when it is generated until the compressors suction it gains temperature no matter how good the isolation of the lines is. Once running, the load can be 50, 75 or 100% and normally they work in parallel with not necessarily the same charge. In high demanding situations, the three compressors work at 100% load at the same time. Once compressed, BOG can be sent out straight (with or without a second compression phase) or mixed with LNG in the re-condenser. This equipment works at 6–10 barg and at those pressures BOG becomes liquid at temperatures up to −130 ºC approximately. Therefore, the temperature gap between LNG coming into the re-condenser and its boiling point is used to cool down the BOG and condensate it, mixing with the LNG which is incorporated to the send out flow. This way BOG is liquefied at very low pressures compared to the send out ones, reducing the energy consumption during the compression and avoiding the economic loss of burning it.

Every compressor has a maintenance program which includes condition based, preventive and predictive maintenance activities. They have overhauls as well, which are scheduled based on the number of running hours. However, the equipment may also suffer failures and have corrective maintenance activities, especially in less

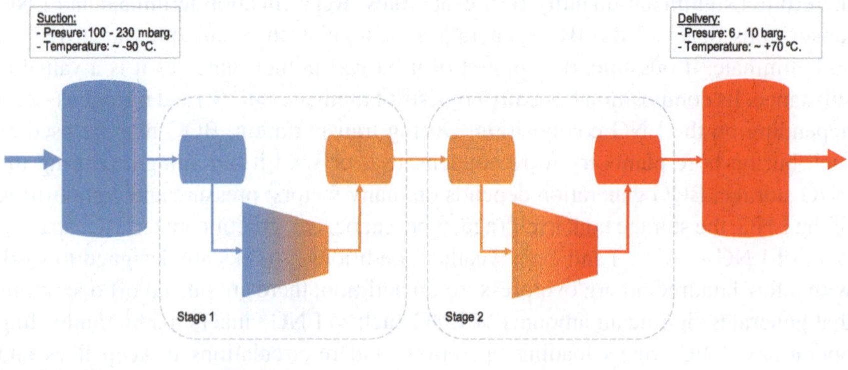

Fig. 7.2 Compressors' basic drawing

critical components, according to current maintenance policy. The most common failure modes that may show up are (in order of importance): rod drop in any of the two stages, gas leak, oil leak, air leak, nitrogen leak, engine failure, high temperature, high pressure, low suction pressure, electrical failure, high vibration…etc. Rod drop failure mode is a consequence of the compressor function. In order to compress the BOG the compressors' piston moves back and forward during the compression cycle. This movement gradually wears down the rider bands leading to a progressive misalignment of the rod known as rod drop. It is important to measure to monitor the rod drop always at the same point of the compression cycle as the rod bends and vibrates.

All measured variables have their importance, and their values may be symptom of a failure. High temperatures may indicate an oil leak as well as the fact that high vibrations may indicate that the compressor has gone out of adjustment. The compressor has a shutdown system which makes a real time on symptom analysis, checking vibrations, and decides if they are acceptable, or the compressor needs to shut down.

There are other variables with the same treatment, but this kind of analysis only includes one variable at each time and does not diagnose the failure mode, it only evaluates one functional condition, and based on manufacturer design and experience decides whether to produce and set off an alarm or shut the equipment down.

The artificial intelligence analysis techniques applied in this research pursue to evaluate several variables at the same time and diagnose one of the most severe failure modes which can present a compressor, the rod drop. This is an evolution of the traditional symptom analysis and the corresponding decision-making process because of the technique and ICT development.

The compressor has the following components: proximity sensors, recycle valve, pressure sensors and transmitters, pressure switches, differential pressure transmitters, temperature sensors and transmitters, accelerometer sensors, dumpers, pressure safety valves and control valves. Rod drop is a failure mode which consist of excessive wearing of the rod, and due its importance, detection, diagnosis, and prognosis have been developed.

Current development now is focused on the detection and diagnosis of the failure, and therefore all the existent knowledge about this failure mode from a proactive point of view is being permanently reviewed. In these reviews, all system variables are collected, structured and validated by expert technicians within failure mode descriptors as in Guillén et al. (2016). The purpose is to determine the more crucial descriptors to detect and diagnose the failure mode beyond traditional methods, searching behaviors patterns through artificial intelligent techniques.

Three cryogenic reciprocating compressors with the same design have been analyzed. Two of them had been on duty for over 28 years and the third for almost 15 years. Initially, a set of eighteen variables directly observed was collected for each compressor at the initial stage of the research. Among them, non-significant features were dismissed as they may be inferred by another variable and only nine variables and two more resulting of the combination of them, having technical meaning, have been finally included.

The selected variables were the more significant ones to determine each compressor's performance and condition. The former is characterized by temperature and pressure at the suction and the discharge, and their calculated respective differences. The latter is related to noise and/or vibrations, running time between failures, load changes between failures, average load between failures and running time pass next maintenance function (see Table 7.1).

In these descriptors, data were sampled hourly as the value of each one at that moment. Even though given the relative stability in the measured variables, since the equipment works in three different regimes, with several load changes a day, filtering of data has been realized carefully with outliers removing and k-NN scattering per regimen. In addition, during the analysis, those data based on demonstrable failures in the measurement elements were eliminated. After that, it was observed that the possible noise effect per regimen had a low effect. In addition, from the CMMS, information of the reported malfunctioning conditions, and overhauls performed on the compressors, were collected, and analyzed. These data were matched with the total running time and the load of each compressor to calculate the following latent variables:

- RT_{BF} Running time between failures (hours).
- L_c Load changes between failures. This variable considers every single change of load in the compressors, including start-ups. Each compressor can run at three different duties.
- L_{avg} Average load between failures. This variable was calculated only during running ours, therefore stand-by hours have not been considered as it would distort the information.

Data has been divided in two different classes: "rop drop" and "normal". The first one has been assigned when the failure mode has been reported and the second one when not. All data were analyzed and filtered removing outliers and datasets with missing information. Table 7.2 shows a summary of the data used. As all the compressors are the same model and process the same fluid (composition, temperature, suction pressure…) they may be studied as one equipment and therefore there are more than 90.000 h of data, equivalent to over ten years of running compressors. Of those, 96% of time correspond to the compressors working on regular load (75 or 100%) and the rest to lower loads and starting-up conditions. This is consistent with the way these compressors are operated.

7.4.2 The Design of a CBM Strategy for the Compressors

There are specific tools for condition monitoring of this type of equipment, but they are not implemented on all of them due to their cost of implementation and management. However, distributed control systems (DCS) have been storing plant data since many years before the birth of the specific tools and are widespread. Computerized

Table 7.1 Symptom analysis and decision-making process

Symptom	Sensor and measurement technique	System variable	Descriptor (unit)	Detection/diagnosis
Running time pass next maintenance function	Timer, routinely check	Compressor running time	RT (h)	Maintenance staff check running time as a routine
Conditions deviation from expected at due load	Load indicator, continuous	Compressor load	L (%)	DCS technician detects it
Noise and/or vibrations	Accelerometer, continuous, DCS and routinely check	Compressor vibration	V ($mm{\cdot}s^{-2}$)	Field technician or DCS technician detects it. DCS alarm goes off
Temperature deviation	Temperature sensor, continuous, DCS and routinely check	Delivery temperature	T_d (C)	DCS technician detects it. DCS alarm goes off
Temperature deviation	Temperature sensor, continuous, DCS and routinely check	Suction temperature	T_s (C)	DCS technician detects it. DCS alarm goes off
Temperature deviation	Temperature sensor, continuous, DCS and routinely check	Suction temperature Delivery temperature	T_d (C)-T_s (C)	DCS technician detects it. DCS alarm goes off
Pressure deviation	Pressure sensor, continuous, DCS and routinely check	Delivery pressure	P_d (barg)	DCS technician detects it. DCS alarm goes off
Pressure deviation	Pressure sensor, continuous, DCS and routinely check	Suction pressure	P_s (barg)	DCS technician detects it. DCS alarm goes off
Differential Pressure deviation	Differential pressure sensor, continuous, DCS and routinely check	Delivery pressure	P_s (barg)-P_d (barg)	DCS technician detects it. DCS alarm goes off

Table 7.2 Data summary for different operating conditions

	Running time (h)					
	Total		Load			
	Hours	%	Other[a]	50	75	100
Compressor 1	32.482	35,75	189	1.617	7.339	23.337
Compressor 2	37.218	40,96	104	1.034	8.497	27.583
Compressor 3	21.156	23,29	37	622	6.079	14.418
Total	90.856		330	3.273	21.915	65.338

[a] "Other" indicates charges under 50% corresponding with starts and stops of the equipment

maintenance management systems, (CMMS) effectively implemented, have information of the maintenance performed on every single equipment and failures that have appeared over the years. If the information of the DCS and the CMMS are crossed, then how a specific equipment has been working and when it has presented a particular failure mode (such as rod drop), can be appreciated.

In addition, to maintain these complex assets, most of the maintenance techniques for planning optimization are based on Reliability-Centered Maintenance (RCM), balancing between preventive and corrective activities [15] and proposing a structured approach that facilitates the extraction of the whole potential of the most important assets in monitored scenarios [17]. To that end, Condition Based Maintenance (CBM) strategies are a common RCM process outcome in compressors, as a key enabler for optimizing asset life-cycle costs and performance [15].

CBM strategies for these compressors can be based on DCS and CMMS data, by using information within these systems, the diagnosis and early prediction of some critical failure modes can be possible. This can be done with enough time to plan and schedule the maintenance activities appropriately, minimizing equipment costs and unavailability. Moreover, the operations duty and equipment conditions may be modulated with the purpose of ensuring that maintenance windows fit the more suitable timing from the process planning point of view.

To design CBM strategies based on DCS, and CMMS compressors data normally requires the identification of the necessary variables per failure mode, in order to detect the incipient failures (point P in the PF curve used in CBM), and then the definition of rules to for real time failure mode prediction when approaching to the functional failure (point F). However, screening among the large number of features available to model the condition of a compressor becomes a serious problem. It may require technical knowledge in data mining and artificial intelligent techniques [2] in addition to the relevant failure physics.

The correct selection of the subset of features may result in models which work faster and produce more accurate predictions, avoiding the curse of the dimensionality phenomenon. To deal with this feature selection problem in compressors failure modes prediction, several feature selection ranking (FSR) methods will be considered. A very important issue of these methods is the fact that the feature selection is done regardless of the Machine Learning algorithm to be applied.

More precisely, this case study was focused on the feature dimensionality problem linked to a rod drop failure mode of the referred LNG compressors and used thirteen FSR techniques to rank features over more than 90.000 combined running hours of three cryogenic reciprocating compressors. Then, three well-known classifiers were used to construct and test the resulting failure prediction model.

7.4.3 The Selection of Features

Different data sets were divided in two test sets (outer and inner) by cross validation over 5 folds to train the feature selection algorithms with the presented classifiers (Fig. 7.3). The validation is done with five external sets that were not used in the developed models.

It is necessary to evaluate the influence that the FSR methods have on the test performance results obtained by classifiers. Besides, as different FSR has different ranked features it is interesting to evaluate the similarity of their results. Even the same FSR may present different rankings depending on the data subsets employed (stability). This was analyzed comparing the ranking results in five folds using same FSR and classifiers.

To evaluate the FSR influence on the classifiers the ranking classification performance curve (RCPC) is used drawing the points $(CM_i(R, H), CM_{i+1}(R, H))$, where CM is the classifier performance metric, being $i \in \{1, \ldots, k\}$ the order in

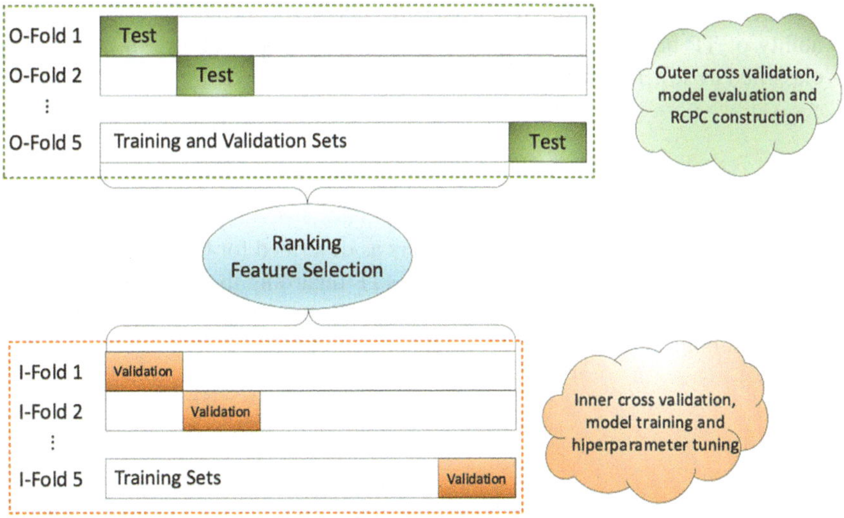

Fig. 7.3 Experimental setup

the ranking, and k is the total feature number. In the RCPC construction four well-known metrics were used: geometric mean, Cohen's kappa, F1 Score and Mathews correlation coefficient. These metrics are defined as follows:

- Cohen's Kappa (Kappa) [16] operates with the points similar to the Receiver Operating Characteristic (ROC) curve, calculating the *Kappa* value (from −1 to 1, showing −1 "total disagreement", 1 "perfect agreement", and next to 0 "random classification") over the following formula:

$$Kappa = \frac{N\sum_{i=1}^{\beta} x_{ii} - \sum_{i=1}^{\beta} x_{i\cdot}\, x_{\cdot i}}{N^2 - \sum_{i=1}^{\beta} x_{i\cdot}\, x_{\cdot i}}$$

  with N samples number, β class values, and with values of the confusion matrix: $x_{i\cdot}, x_{\cdot i}, x_{ii}$ are respectively the total counts of rows, columns, and the main diagonal.
- Matthews correlation coefficient (MCC): this metric is equivalent to the Pearson phi coefficient, which means that MCC is a correlation coefficient for the prediction method regarding the observed data. It can be calculated based on:

$$MCC = \frac{TP \cdot TN - FP \cdot FN}{\sqrt{(TP+FP)\cdot(TP+FN)\cdot(TN+FP)\cdot(TN+FN)}}$$

  with the same definition of the parameters as described for Cohen's kappa. It has some advantages over F_1 Score as MCC considers the four classes of the confusion matrix in its results.
- Geometric mean (G-Mean): it considers the geometric mean of sensitivity (probability of TP being classified as so) and specificity (probability of a TN being classified as so). This metric is defined as follows:

$$G - Mean = \sqrt{\frac{TP}{TP+FN} \cdot \frac{TN}{TN+FP}}$$

  with the same definition of the parameters as described for Cohen's kappa.
- F_1 Score: this metric considers precision (TP regarding the number of positives predicted, both TP and FP) and the recall (TP regarding the number of actual positives, both TP and FN). It is calculated as follows:

$$F_1 Score = \frac{2 \cdot TP}{2 \cdot TP + FP + FN}$$

  with the same definition of the parameters as described for Cohen's kappa. Its main disadvantage is that it gives the same weight to precision and to recall when different mistakes may have different consequences.

7.4.4 *The Results of the Analysis*

The results of the comparative study are organized as follows:

1. Classifiers are analyzed to see differences in their performance, to test whether all of them are suitable to deal with the problem or not.
2. FSR result are reviewed to see if the value of the kappas reached is good enough to be considered as appropriate for this field of study, while the other three metrics are analyzed, and their results compared with the ones obtained with the kappas.
3. A set of FSR are selected and the top-5 ranked features for each one is compared for the five performed folds to know if each FSR are repeatable enough and independent on the subset of data chosen for the training stage.
4. All FSR top-5 ranked variables are compared to see if there are certain features which have in general more importance than others and determine the stability of the FSR.
5. A technical review of the results of the previous analysis is carried out to confront theoretical findings with practical evidence and experience.

In Fig. 7.4, as an example, results for Cohen's kappa metric are presented. It can be appreciated how k-NN and RF had very similar values of kappa for all the FSR methods, with the latter performing slightly better. However, NB presented poor results compared with the other two algorithms and for all the FSR methods. NB behaved poorly with several FSR and, with most of them, results were even worse when the number of features considered was increased. This behavior was observed during the five folds performed. Therefore, NB was discarded.

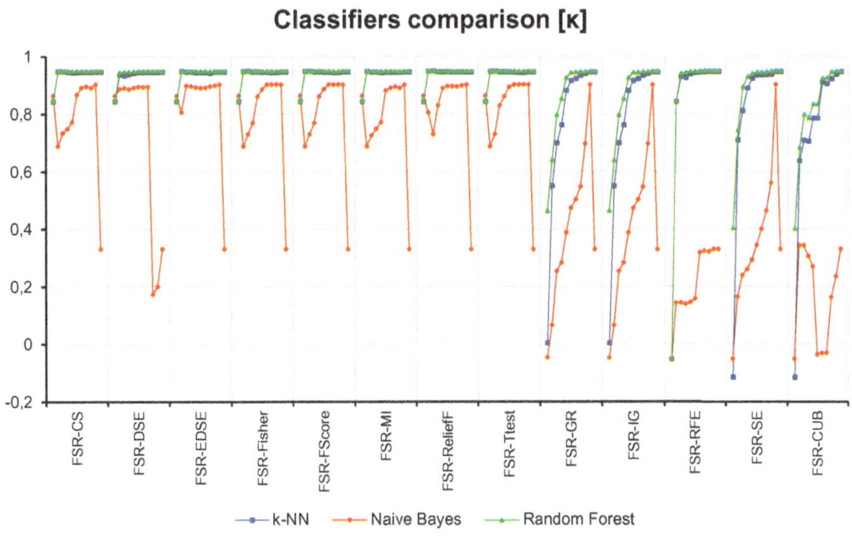

Fig. 7.4 Classifiers comparison during the five folds (Cohen's kappa)

Figure 7.4 also shows that FSR GR, IG, RFE, SE and CUB performance were worse than the rest of the methods with k-NN and RF algorithm as they needed a higher number of features to reach acceptable values of kappa, which the other FSR obtained with one or two features. The referred five FSR methods were rejected. As it happened with the classifiers, the behavior observed was the same in the five folds performed.

Once the classifiers and FSR were selected, Fig. 7.5 shows the kappa values obtained for all of them by the top-5 ranked variables (similar results were observed for the five folds). The performance was quite similar for all the FSR and classifiers with light differences which may be considered naive, being the combination of K-NN and DSE the worse. In general, k-NN algorithm had decreasing kappa values as the number of features increased.

Once FSR and classifiers have been selected and validated regarding the values of kappa obtained, the repeatability and stability of the entire process may be analyzed. As one of the objectives pursued is to reduce the number of features needed to diagnose the rod drop failure mode appearance, only the top-5 ranked features were considered. To evaluate the repeatability, it must be studied if the features selected within the first five positions were the same for each FSR method individually for the five folds and if they were selected at the same position regardless the fold. To analyze the stability, all the FSR were compared to determine which features were most frequently ranked and their position.

As can be seen in Table 7.3, the FSR CS and Ttest were repeatable as they ranked the same features in the same position for the five folds. However, each method gave its own selection, although they agreed on three features giving them the same position. Table 7.3 shows the K-NN results for MCC and F1 Score results. Fisher,

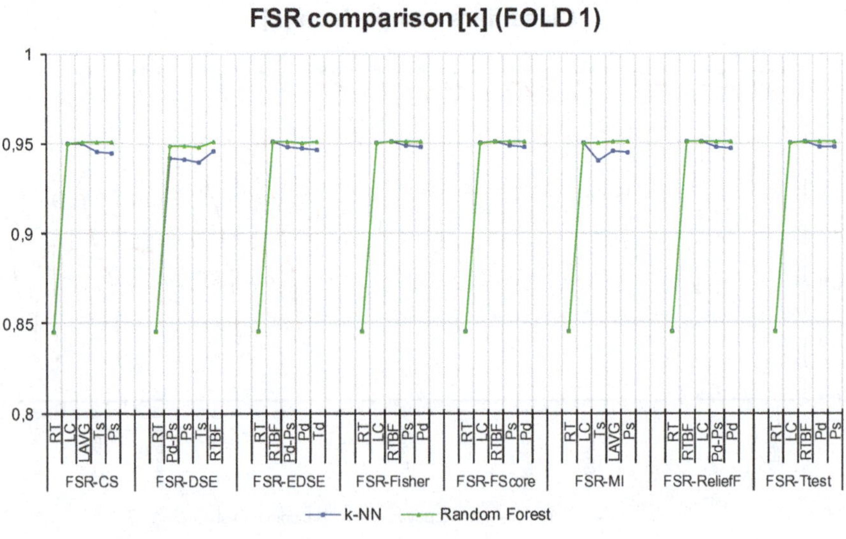

Fig. 7.5 FSR comparison during the five folds

Table 7.3 Features ranked in each fold and FSR

FSR	Features: Fold 0	Fold 1	Fold 2	Fold 3	Fold 4
FSR-CS	RT (0,887; 0,861)	RT (0,89; 0,864)	RT (0,891; 0,866)	RT (0,891; 0,866)	RT (0,889; 0,864)
	L_C (0,965; 0,965)	L_C (0,966; 0,966)	L_C (0,966; 0,966)	L_C (0,965; 0,965)	L_C (0,965; 0,965)
	L_{AVG} (0,966; 0,965)	L_{AVG} (0,965; 0,965)	L_{AVG} (0,966; 0,966)	L_{AVG} (0,965; 0,964)	L_{AVG} (0,966; 0,965)
	T_s (0,964; 0,964)	T_s (0,962; 0,961)	T_s (0,963; 0,963)	T_s (0,963; 0,963)	T_s (0,963; 0,963)
	P_s (0,963; 0,961)	P_s (0,962; 0,960)	P_s (0,962; 0,961)	P_s (0,962; 0,961)	P_s (0,962; 0,961)
FSR-Ttest	RT (0,887; 0,861)	RT (0,89; 0,864)	RT (0,891; 0,866)	RT (0,891; 0,866)	RT (0,889; 0,864)
	L_C (0,965; 0,965)	L_C (0,966; 0,966)	L_C (0,966; 0,966)	L_C (0,965; 0,965)	L_C (0,965; 0,965)
	RT_{BF} (0,966; 0,966)	RT_{BF} (0,966; 0,966)	RT_{BF} (0,966; 0,966)	RT_{BF} (0,965; 0,965)	RT_{BF} (0,966; 0,966)
	P_d (0,964; 0,964)	P_d (0,963; 0,963)	P_d (0,965; 0,964)	P_d (0,965; 0,965)	P_d (0,964; 0,964)
	P_s (0,965; 0,964)	P_s (0,964; 0,963)	P_s (0,964; 0,963)	P_s (0,964; 0,963)	P_s (0,964; 0,964)
FSR-Fisher	RT (0,887; 0,861)	RT (0,89; 0,864)	RT (0,891; 0,866)	RT (0,891; 0,866)	RT (0,889; 0,864)
	L_C (0,965; 0,965)	L_C (0,966; 0,966)	L_C (0,966; 0,966)	L_C (0,965; 0,965)	L_C (0,965; 0,965)
	RT_{BF} (0,966; 0,966)	RT_{BF} (0,966; 0,966)	RT_{BF} (0,966; 0,966)	RT_{BF} (0,965; 0,965)	RT_{BF} (0,966; 0,966)
	P_s (0,964; 0,964)	P_s (0,964; 0,964)	P_d (0,965; 0,964)	P_s (0,964; 0,963)	P_s (0,964; 0,963)
	P_d (0,965; 0,964)	P_d (0,964; 0,963)	P_s (0,964; 0,963)	P_d (0,964; 0,963)	P_d (0,964; 0,964)
FSR-FScore	RT (0,887; 0,861)	RT (0,89; 0,864)	RT (0,891; 0,866)	RT (0,891; 0,866)	RT (0,889; 0,864)
	L_C (0,965; 0,965)	L_C (0,966; 0,966)	L_C (0,966; 0,966)	L_C (0,965; 0,965)	L_C (0,965; 0,965)
	RT_{BF} (0,966; 0,966)	RT_{BF} (0,966; 0,966)	RT_{BF} (0,966; 0,966)	RT_{BF} (0,965; 0,965)	RT_{BF} (0,966; 0,966)
	P_s (0,964; 0,964)	P_s (0,964; 0,964)	P_d (0,965; 0,964)	P_s (0,964; 0,963)	P_s (0,964; 0,963)
	P_d (0,965; 0,964)	P_d (0,964; 0,963)	P_s (0,964; 0,963)	P_d (0,964; 0,963)	P_d (0,964; 0,964)
FSR-MI	RT (0,887; 0,861)	RT (0,89; 0,864)	RT (0,891; 0,866)	RT (0,891; 0,866)	RT (0,889; 0,864)
	L_C (0,965; 0,965)	L_C (0,966; 0,966)	L_C (0,966; 0,966)	L_C (0,965; 0,965)	L_C (0,965; 0,965)
	L_{AVG} (0,966; 0,965)	T_s (0,958; 0,955)	L_{AVG} (0,966; 0,966)	L_{AVG} (0,965; 0,964)	L_{AVG} (0,966; 0,965)
	T_s (0,964; 0,964)	L_{AVG} (0,962; 0,961)	T_s (0,963; 0,963)	T_s (0,963; 0,963)	T_s (0,963; 0,963)
	P_s (0,963; 0,961)	P_s (0,962; 0,960)	P_s (0,962; 0,961)	P_s (0,962; 0,961)	P_s (0,962; 0,961)
FSR-ReliefF	RT (0,887; 0,861)	RT (0,89; 0,864)	RT (0,891; 0,866)	RT (0,891; 0,866)	RT (0,889; 0,864)
	RT_{BF} (0,966; 0,966)	RT_{BF} (0,966; 0,966)	RT_{BF} (0,966; 0,966)	RT_{BF} (0,965; 0,965)	RT_{BF} (0,966; 0,966)
	L_C (0,966; 0,966)	L_C (0,966; 0,966)	L_C (0,966; 0,966)	L_C (0,965; 0,965)	L_C (0,966; 0,966)
	P_d (0,964; 0,964)	P_d-P_s (0,964; 0,963)	P_d (0,965; 0,964)	P_d (0,965; 0,965)	P_d (0,964; 0,964)
	P_d-P_s (0,964; 0,963)	P_d (0,963; 0,962)	P_d-P_s (0,964; 0,963)	P_d-P_s (0,964; 0,964)	P_d-P_s (0,964; 0,963)
FSR-EDSE	RT (0,887; 0,861)	RT (0,89; 0,864)	RT (0,891; 0,866)	RT (0,891; 0,866)	RT (0,889; 0,864)
	RT_{BF} (0,966; 0,966)	RT_{BF} (0,966; 0,966)	RT_{BF} (0,966; 0,966)	RT_{BF} (0,965; 0,965)	RT_{BF} (0,966; 0,966)
	P_d (0,964; 0,964)	P_d-P_s (0,964; 0,963)	P_d (0,964; 0,964)	P_d (0,965; 0,964)	P_d (0,964; 0,964)
	P_d-P_s (0,964; 0,963)	P_d (0,963; 0,963)	P_d-P_s (0,964; 0,964)	P_d-P_s (0,964; 0,963)	P_d-P_s (0,964; 0,963)
	T_d (0,962; 0,961)	T_d (0,962; 0,961)	T_d (0,962; 0,961)	T_d (0,963; 0,962)	T_d (0,962; 0,961)
FSR-DSE	RT (0,887; 0,861)	RT (0,890; 0,864)	RT (0,891; 0,866)	RT (0,891; 0,866)	RT (0,889; 0,864)
	P_d-P_s (0,961; 0,960)	P_d-P_s (0,959; 0,957)	P_d-P_s (0,962; 0,960)	T_s (0,950; 0,945)	T_s (0,949; 0,943)
	T_s (0,956; 0,953)	P_s (0,958; 0,956)	T_s (0,955; 0,951)	P_d-P_s (0,955; 0,952)	P_s (0,946; 0,939)
	P_s (0,958; 0,955)	T_s (0,957; 0,955)	P_s (0,956; 0,953)	P_s (0,958; 0,955)	P_d-P_s (0,957; 0,954)
	P_d (0,957; 0,955)	RT_{BF} (0,962; 0,961)	L_{AVG} (0,961; 0,960)	RT_{BF} (0,961; 0,959)	L_{AVG} (0,963; 0,961)

(a; b): K-NN results for F_1 Score (*a*) and for MCC (*b*).

White background: feature ranked with the same position in all folds.

Pale grey background: feature ranked in all folds within the top-5 important ones, but not with the same position.

Dark grey background: feature not ranked in all folds.

FScore, MI, ReliefF y EDSE methods were repeatable regarding the features selected during the five folds as well. In these cases, the order of some features varied from a fold to another and therefore these FSR were considered less repeatable than CS and Ttest.

Finally, the FSR DSE, which had shown a slightly poorer performance when analyzing the values of kappa, only repeated four features during the five folds, keeping the same position only for one of them. There are three features that were ranked at fifth place or excluded from the selection depending on the fold. This seemingly lack of repeatability for some of the FSR may be justified considering that, as the number of used features increases, the kappa values varied less than a thousandth from one featured to the next ranked and at highest it varied no more than 3 thousandths. With this order of magnitude, the change of one position in the ranking is very plausible. As the results shown, the two best ranked variables always kept their position in almost all the FSR but one, and most of the times the third one did the same. Therefore, it may be assumed that the best three ranked variables had a stronger importance than the others. In general, the FSR results may be considered repeatable, with the DSE the only method which may be discarded.

Only nine out of the original eleven variables were ranked among the top-5 features (see Table 7.4). The FSR methods Ttest, Fisher and FScore ranked the very same variables and almost in the same order, in fact the latter two kept the same order in the five folds. ReliefF shared four of the features ranked with them, the first three variables repeated although not in the same order.

In general, the methods with the greater number of features in common tended to rank the same variables among the best three. Table 7.4 was arranged considering the features more repeated from left to right. The order of the features seemed to be the same as if the mode of the ranked position and the repeatability would have been considered.

RT was always ranked at first for all FSR and in all the folds, being therefore identified as the most important one.

In five of the eight FSR, L_C was ranked in second place and in third place in the sixth method. In two FSR was not ranked, the ones with less in common with the other methods, EDSE and DSE.

RT_{BF} was ranked in third place by three FSR, in second by another two and in fifth by another one. Two methods did not rank RT_{BF} among the top-5 features.

Table 7.4 Resume of the features ranked and the FSR method

FSR	RT	L_C	RT_{BF}	P_s	P_d	L_{AVG}	T_s	P_d-P_s	T_d
FSR-Ttest	1	2	3	5	4				
FSR-Fisher	1	2	3	4-5	4-5				
FSR-FScore	1	2	3	4-5	4-5				
FSR-MI	1	2		5		3-4	3-4		
FSR-CS	1	2		5		3	4		
FSR-ReliefF	1	3	2		4-5			4-5	
FSR-EDSE	1		2		3-4			3-4	5
FSR-DSE	1		5	2-3-4	5	5	2-3-4	2-3-4	
Times ranked	8	6	6	6	6	3	3	3	1

White background: feature ranked with the same position in all folds.
Pale grey background: feature ranked in all folds within the top-5 important ones, but not with the same position.
Dark grey background: feature not ranked in all folds.

Ps and Pd were ranked in fourth or fifth position by six and five FSR respectively, although only three methods ranked both together. EDSE ranked only P_d but in third and fourth place. DSE ranked P_s in second, third and fourth position, but P_d was ranked fifth in one of the five folds and left out in the other four.

Finally, L_{AVG}, T_S and P_d-P_s, were ranked by only three methods, generally with a variable position depending on the fold, and T_d was ranked only by EDSE and constantly in fifth place. L_{AVG} and T_S were ranked in third or fourth position by MI and CS, T_s even second or third by DSE, but this method only ranked L_{AVG} in fifth position in two folds and did not select it in the other three. P_d-P_s was ranked by ReliefF fourth and fifth, by EDSE third or fourth and by DSE second, third or fourth.

If the lack of repeatability shown by FSR-DSE results were considered and it was discarded, Table 7.4 would keep the same order.

Although the FSR GR, IG, RFE, SE and CUB were rejected because of their low kappa scores, their repeatability and stability were also briefly analyzed. As expected, these FSR had poor repeatability and the features selected were not the same in all folds. In general, the eleven features were selected at least once. None of the FSR were able to keep the same variables in the same position in the five folds. Only CUB ranked the same top-5 variables in every fold and RFE behaved like DSE which was not discarded originally. SE kept only three variables in all the folds, whereas GR and IG ranked the very same features at the same positions, but only one variable was selected in all the folds but placed differently. Regarding the selected features, the ranking order was almost the opposite of the one observed with the best FSR. Only L_{AVG} was ranked quite similar. V was selected only by RFE but it was surprisingly ranked first in all the folds. T_d-T_s was ranked by four FSR from the first to the fifth position depending on the fold and method. With these observations, the rejection of these FSR was right and DSE could be considered as arguably, just like it had proven all throughout the analysis.

7.4.5 *Discussion of Results*

In general, the results seemed logical and coherent. Total running time (RT) represents the ageing of the equipment. The more RT the closer to the end their lifetime and the more severe wear of components and eventually the higher probability of a rod drop failure mode appearance. Overhauls performed on the compressors pursue the lengthening of the lifetime preventing the appearance of failure modes, although these maintenance actions are not technically and economically viable to prevent all possible failure modes and are arranged with the integrity of the whole equipment in mind.

Changes in load (L_c) generate instabilities which lead to a more severe wear down as well. It must be remembered that every start-up of an equipment, which is no more than an extreme change of load, in general counts as 100–130 running hours. As a matter of fact, start-ups have been considered as load changes in this research. Therefore, load changes may be converted into running hours. This feature is less important than the total running time because this equipment works almost on a non-stop basis.

Running time between failures (RT_{BF}) analysis is analogue to the running time one. After repairing an equipment to remove a failure mode it will appear again after a certain number of running hours. This feature importance compared to total running time and load changes may be inferior to the weight of ageing and instabilities in the equipment health. Additionally, the running time between failures may be considered as somehow included in the total running time.

Suction and delivery pressure, as well as the corresponding temperatures, force the equipment to work in a more severe or light demanding way and therefore affect to its wear. It is not odd to find them among the most influent features, with a relative less weight than the running time related features and the load changes, which is logical as well, considering that this equipment works mostly with steady conditions once they have started up. Pressure variations are more frequent and severe than the temperature ones, so it is acceptable their importance to be higher.

Average load (L_{AVG}) is poorly ranked which may be surprising, however it must be considered that once the steady running condition is reached the equipment suffer less wear compared with the variations in other parameters, which additionally are more frequent such as in the case of pressures and temperatures.

Another important result is the fact that vibrations were not ranked by any of the methods among the top-5 features. It can be thought that the more vibrations the more wear the equipment would suffer and as vibration increases with the lifetime of the equipment, unless specific actions to reduce it are taken, it must be a more influent feature. This lack of relevance could be explained by the following facts:

- This equipment has a shutdown trigger by vibrations. If they pass a setting point an alarm goes off and a maintenance notice is produced. If they increase to dangerous levels the equipment shuts down.
- It is a feature with a special monitoring to keep it within acceptable values. Other variables are paid less attention and they are changed, if necessary (load and delivery pressure), vary naturally according to the process conditions (suction pressure, suction, and delivery temperature) or are unavoidable (running hours).
- In this research vibrations have been analyzed as hourly samples, perhaps a comparative analysis of them would gain weight to this feature.

Finally, the failure of a compressor when the plant has a peak of BOG generation could lead to a situation in which the pressure of the tanks rises to dangerous levels and therefore part of the BOG is flared. The energy burnt in a flare is a dolorous inefficiency since all the cost from the well to the regasification plant is lost, the environmental impact does not have a benefit for the society and profits from the sale of that energy are missed. Therefore, the risk of unavailability of a compressor and the costs associated to it can be reduced following this method for failure mode identification, as well as the maintenance and warehouse associated costs.

7.5 Conclusions

The curse of dimensionality is a phenomenon that data scientist must consider. Machine Learning models performance may be importantly impacted by this problem when dealing with a relevant number of features.

This Chapters includes a case study where a process for features reduction is followed, the process considered a reasonable set of FSRs and Classifiers. This process can now be repeated for similar equipment with the same tools in a very efficient way.

Reducing the number of variables monitored to predict failure modes would also result in less data storage space and analysis efforts, which reduces the investment in hardware, software, and labor costs.

Regarding maintenance activities, they can be planned and scheduled with great timing according to the process needs. Thanks to this, the spare parts management can be done in advance properly and the mean time to repair (MTTR) reduced. Furthermore, it will allow the avoidance of other major failures as a premature failure (rod drop in the case study) may be indicative of other problems which could be affecting different parts of the asset so other failure modes would appear soon. By monitoring the evolution of the predictive parameters associated to the diagnostic, the Reliability department or team could coordinate with Operations to suggest changes in the priority of use for the different assets (compressors in the case study) to reach maintenance windows at the most suitable moment for the facility needs.

References

1. Koutroumbas K, Theodoridis S (2008) Pattern recognition, 4th ed. Burlington. ISBN 978-1-59749-272-0. Retrieved 8 January 2018
2. Gray CS, Watson SJ (2010) Physics of failure approach to wind turbine condition-based maintenance. Wind Energy 13(5):395–405. https://doi.org/10.1002/we.360
3. García-Pedrajas N, del Castillo JAR, Cerruela-García G (2015) A proposal for local $ k values for k-nearest neighbor rule. IEEE Trans Neural Netw Learn Syst 28(2):470–475. https://doi.org/10.1109/TNNLS.2015.2506821
4. Liu Y (2004) A comparative study on feature selection methods for drug discovery. J Chem Inf Comput Sci 44(5):1823–1828. https://doi.org/10.1021/ci049875d
5. Godden JW, Bajorath J (2003) An information-theoretic approach to descriptor selection for database profiling and QSAR modeling. QSAR Comb Sci 22(5):487–497. https://doi.org/10.1002/qsar.200310001
6. Dash M, Choi K, Scheuermann P, Huan L (2002) Feature selection for clustering-a filter solution. In: IEEE international conference on data mining. Proceedings, Maebashi City, Japan, pp 115–122. https://doi.org/10.1109/ICDM.2002.1183893
7. Elisseeff A, Schölkopf B, Pérez-Cruz F, Weston J, Bousquet O, Chapelle O (2003) Feature selection and transduction for prediction of molecular bioactivity for drug design. Bioinformatics 19(6):764–771. https://doi.org/10.1093/bioinformatics/btg054

8. Hemmateenejad B, Mehdipour A, Deeb O, Sanchooli M, Miri R (2011) Toward an optimal approach for variable selection in counter-propagation neural networks: modeling protein-tyrosine kinase inhibitory of flavanoids using substituent electronic descriptors. Mol Inf 30(11–12):939–949. https://doi.org/10.1002/minf.201100081
9. Zhou L, Lai KK, Yen J (2012) Empirical models based on features ranking techniques for corporate financial distress prediction. Comput Math Appl 64(8):2484–2496. https://doi.org/10.1016/j.camwa.2012.06.003
10. Alhaj TA, Siraj MM, Zainal A, Elshoush HT, Elhaj F (2016) Feature selection using information gain for improved structural-based alert correlation. PLoS One 11(11):e0166017. https://doi.org/10.1371/journal.pone.0166017
11. Vergara JR, Estévez PA (2014) A review of feature selection methods based on mutual information. Neural Comput Appl 24(1):175–186. https://doi.org/10.1007/s00521-013-1368-0
12. Kononenko I (1994) Estimating attributes: analysis and extensions of RELIEF. In: European conference on machine learning. Springer, pp 171–182. https://doi.org/10.1007/3-540-57868-4_57
13. Guyon I, Weston J, Barnhill S, Vapnik V (2002) Gene selection for cancer classification using support vector machines. Mach Learn 46(1–3):389–422. https://doi.org/10.1023/A:1012487302797
14. Pedregosa F, Varoquaux G, Gramfort A, Michel V, Thirion B, Grisel O, Blondel M, Prettenhofer P, Weiss R, Dubourg V, Vanderplas J, Passos A, Cournapeau D, Brucher M, Perrot M, Duchesnay E (2011) Scikit-learn: Machine learning in Python. J Mach Learn Res 12:2825–2830
15. Crespo Márquez A (2007) The maintenance management framework. Models and methods for complex systems maintenance. Springer, UK. ISBN 978-1-84628-821-0, https://doi.org/10.1007/978-1-84628-821-0
16. Ben-David A (2007) A lot of randomness is hiding in accuracy. Eng Appl Artif Intell 20(7):875–885. https://doi.org/10.1016/j.engappai.2007.01.001
17. Higgins LR, Mobley RK, Smith R (2002) Maintenance engineering handbook, 6th ed. McGraw-Hill, New York. https://doi.org/10.1036/0071394524
18. Magdy S (2000) The curse of dimensionality. IME Company. http://www.infme.com/curse-of-dimensionality-ml-big-data-ml-optimization-pca/
19. Contreras-Valdes A, Amezquita-Sanchez JP, Granados-Lieberman D, Valtierra-Rodriguez M (2020) Predictive data mining techniques for fault diagnosis of electric equipment: a review. Appl Sci 10(3):950

Chapter 8
The Dynamic Measurement of Failure Risk

8.1 Introduction to the Problem

Within high digitalization operational contexts, a great amount of information related to assets operation and condition emerges. Linking this information to a reasonable measure of failure risk becomes critical for the process of CBM decision making.

A good measure of failure risk allows gaining control for asset functionality preservation, avoiding undesirable failure consequences effectively and efficiently. To take advantage of digitization requires a more sophisticated way of scheduling and carrying out maintenance activities, relying on a very dynamic assessment of failures risk [1]. Digital Transformation (DT) must support a more powerful and effective failure risk management (RM). DT provides more data, information and knowledge of failure that should be converted into the continuous improvement of dynamic failure risk control for a real intelligent maintenance decision making. The idea of linking maintenance management with risk management is well presented in ISO 55000. Risk management is explicitly introduced in asset management (AM) principles [2, 3] which presents maintenance as the principal tool for engineering AM.

According to this context within intelligent maintenance and asset management evolution, three main gaps can be identified in many industrial and infrastructure firms:

- *Low capacity for assessing assets dynamic risk level.* Towards this aim, how to translate the great amount of information that is currently available, into simple ways to interpret and control risk, is still a gap for many industrial practitioners. This will introduce important advantages for maintenance management optimization and risk control over time. In this paper, risk level states are proposed as key elements supporting intelligent maintenance: (i) for interpreting real-time data on the system evolution in terms of risk level; (ii) as a trigger of maintenance decision-making, that, in addition produces effects on risk level (reduction or contention) (iii) for defining and recording events, which are fundamental for risk evolution control.

A. Crespo Márquez, *Digital Maintenance Management*, Springer Series in Reliability Engineering, https://doi.org/10.1007/978-3-030-97660-6_8

- *Difficult interpretation of risk results by the maintenance planners*, connecting these results to the CBM decision-making processes.
- *Lack of dynamic maintenance activities scheduling*. The massive use of CBM generates dynamic and highly complex decision making. Not only the information available but the number of events and alarms, many of them interconnected. The handling of all these complex scenarios is a barrier for the practical implementation of CBM solutions within preventive maintenance plans. How to facilitate the compression of the information provided by these solutions and the simple interconnection with the maintenance actions scheduling processes, which include human participation, continues to be an unsolved scientific-technical problem of great industrial impact [4].

Risk management (RM) is a very complex activity that assumes a very high responsibility. Safety management systems are based on risk assessment methods, and safety-related events have legal effects on the organization and its staff. Thus, organizations are totally reluctant to change their RM or to introduce RA (Risk Assessment) modifications. But digital transformation reimagines risk management [1, 5]. There are elements that were not possible before or are greatly enhanced by digital transformation. For example, advance CBM, which is also named PHM/CBM or CBM+ [6], allows enhancing transparency to assets' capabilities and their current conditions in order to discover hidden patterns that anticipate risky scenarios [7, 8]. A critical aspect, in this sense, is how to interpret in terms of risk the information generated by new applications based on technologies such as predictive analytics; in other words, how to connect the new information and knowledge available, over time, with effective risk control [1]. DT will provide more information and knowledge of risk and the real chance, not available until now, to monitor risk evolution, allowing for a robust, reliable, and safe transformation from risk assessment (RA) to dynamic risk assessment (DRA).

Owing to the extensive growth of condition based and predictive maintenance (CBM and PdM) references (from academia and industry), that generated a non-depreciable noise over these terms definition and use, a brief description of the interpretation of them is needed in this work. We use as the main reference the international standard CEN-EN 13306_Maintenance Terminology [9]. Here CBM is a type of preventive maintenance (PM), and PdM is a type of CBM. This way, when we refer to CBM in the text, PdM is included. In the same way, predictive analytics (PdA) or prognosis health management (PHM) techniques can be considered as CBM supporting techniques [10]. In the CBM process, data analysis allows detection, diagnosis and prognosis, and these three results can be used as basic inputs for maintenance planning, the final goal of the CBM process [11, 12].

8.2 Dynamic Risk Level Assessment in Literature

Dynamic nature of risk assumes that risk is changing continuously. Over the last few years, interest in the study of dynamic risk has grown substantially as shown in Figure.

This Figure presents number of papers including keywords in Table 8.1, in their titles, abstracts, and/or keywords in the referred well-known journals, and over the years 2015–2020. This search resulted in 262 papers. From the resulting set, we kept papers that dealt with complex systems and had one or more of the following elements: a system-level perspective, systems logic modeling, use of different information sources, and dynamic/online/real-time assessment. In view of Table 8.1, the study of failure modes has been present in the study of risk, however, the study of the condition to measure dynamic risk has been discussed little in the relevant literature. In the same way, there appears to be little relationship between maintenance decision making and dynamic risk assessment.

Currently, there is a need to work with DRA, especially for maintenance, but there is no widely accepted method [5]. This fact confirms the relevance of this Chapter. The main challenges of dynamic risk assessment are related to the improvement in the accuracy of the models and to the ability to update the assessment dynamically [13].

There are many quantitative models to assess risk in one way or another. However, the mathematical models used to quantify risk introduce high uncertainty. In addition, according to [1], it is increasingly difficult to quantify risk, which causes the models to evolve into complex mathematical formats that in turn make decision making complicated.

Although most of the DRA-related literature does not cover decision-making, the results of the risk assessment should allow decision-makers to meet their objectives and understand the uncertainty associated with risk. The goal must be to facilitate decision-making in maintenance according to risk. This has been studied in the literature, although with a different focus, for instance in [14] the objective is to optimize the long-term maintenance plan and optimize the maintenance cost by maximizing the RUL. Others contributions, for example [4], focus on maintenance strategy/policy rather than decision-making in the operational phase. In [17], unlike

Table 8.1 Selected search keywords for finding the most relevant available literature (2015–2020)

Selected search keywords	RESS	Safety science	Computers in industry
Dynamic and risk	88	74	5
Dynamic risk and CBM	5	5	0
Risk and (FMEA OR failure mode)	30	23	0
Dynamic risk and maintenance	14	3	1
Dynamic risk and (PRA OR QRA)	10	4	0

conventional approaches that restrict all maintenance activities to a finite planning horizon, they focus on activity-to-activity scheduling without specifying the horizon.

In addition to decision-making, it's important to integrate the CBM and the information from maintenance activities, so that it collects all the available information, and the risk calculation is as accurate as possible. This integration has been studied in a few cases in the relevant literature. In some cases, as [18] make a distinction between the static view of the fault distributions derived from historical data and the dynamic view of the remaining life derived from the condition, trying to integrate the PHM into the DRA. On the other hand, in more recent studies [15] propose a framework to integrate the CBM and the inspection data in the DRA. Condition monitoring data is collected online by sensors and is indirectly related to component degradation. Inspection data is recorded in physical inspections that directly measure component degradation. However, in this study they do not incorporate the failure modes of the components. They will be incorporated into the proposed model presented later.

8.3 Risk Management (RM), Assessment (RA) and Analysis (RAn)

ISO 31000 defines RM as coordinated activities to direct and control an organization regarding risk. Aven et al. [16] argue that RM is about balancing different concerns, profits, safety, reputation, etc. (value centered elements); and considering a set of alternatives, evaluates their pros and cons, and makes a decision that best meets the decision-makers' values and priorities [16]. The ISO standard introduces a general risk management framework (Fig. 8.1). By going through this scheme completely (from top to bottom and from left to right) it becomes explicit how the RM, as a process, begins with value, or is based on value, and ends in the practical treatment of risks.

This way RA should be interpreted as a part of RM. RA aims to provide risk information to support decision making, structuring the information and knowledge available by providing systemic modeling and understanding, from components level

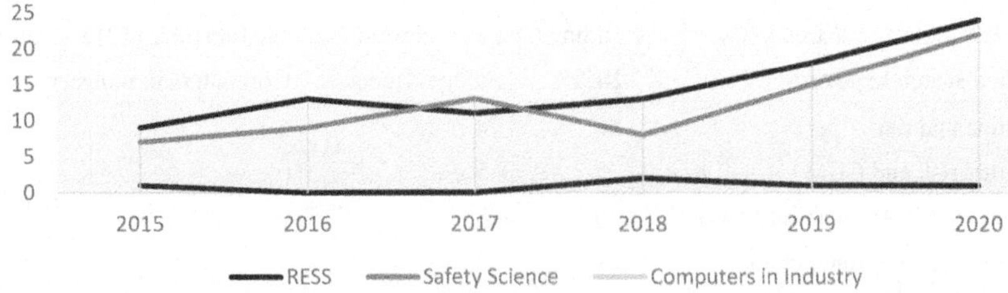

Fig. 8.1 Literature available with Dynamic Risk keywords for the period (2015–2020)

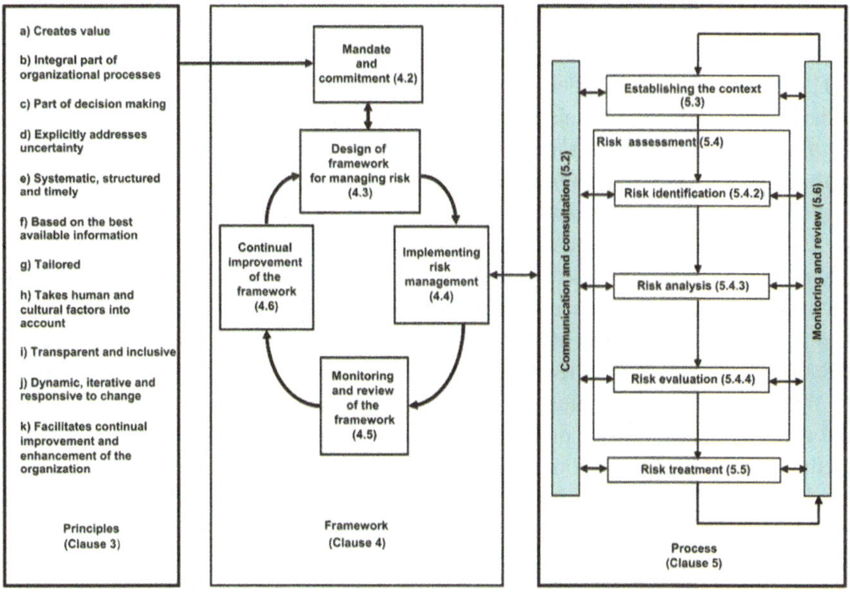

Fig. 8.2 Risk management framework by ISO 31000

to system level [24]. To do that it is necessary to quantify risk and, if it is possible, to introduce probability for managing uncertainty related to these quantities [17, 18].

The idea of quantification of risk is reinforced in QRA concept (Quantitative Risk Assessment). As defined by NORSOK Standard Z-013 [19] and by ISO/IEC standard [19]. QRA includes establishment of the context, risk identification, performance of the risk analysis, risk evaluation. There has been a move from classical deterministic RA or QRA towards PRA (probabilistic risk assessment) integrated approaches that combines the insights provided by the deterministic approach and those from the probabilistic approach with any other requirements in making decisions, generating an adequate framework within which expert opinions can be combined with statistical data to provide quantitative measures of risk [12]. And the last step in this evolution is to develop DRA and integrate with PRA, which is the most suitable approach to take advantage of new digitalization context and to provide RA tools for digitalization context requirements (see Fig. 8.2).

8.4 Introduction to Dynamic Risk Assessment (DRA)

DRA requires accurate knowledge about systems behavior, including their risk evolution, in real time [20, 21]. It is assumed that risk can be accurately quantified and updated using digital capabilities and tools. In particular, it is easy to see the link between monitoring and CBM with risk evolution control [7, 22]. But CBM not only provides dynamic information of risk level, mainly and even before this, allowing the description in great detail of the failure mechanism and failure mode evolution [11]. In this way, the CBM design process is able to be presented as a powerful tool in order to deal with one of main risk evolution challenges elicited by SRA: "*How can we describe and represent the results of risk assessment in a way useful to decision makers, which clearly presents the assumptions made and their justification with respect to the knowledge which the assessment is based upon?*" [23].

Zeng and Zio [12] propose to use the DRA definition previously introduced by Khan et al. [24]: "RA method that updates the estimated risk of a deteriorating process according to the performance of the control system, safety barriers, inspection, and maintenance activities, the human factors, and the implementation of procedures". DRA is suitable to be applied to all asset life cycle phases but with special impact on the operation and maintenance phase, supporting a great improvement in risk-informed decision-making processes [25]. Zeng and Zio in their work introduce a method to mix PRA with condition monitoring results. Villa et al. [13] analyzed the need for a way forward in risk assessment, from QRA to dynamic risk assessment approaches. These authors present a specific approach called the "risk barometer", which aims to provide operators with a risk-based decision support tool. This tool combines three sources of information in the representation of risk: real-time information on safety barriers, the results of risk analyses and expert opinions. Risk barometer experience [17] main points: (i) a previous QRA is required; (ii) the use of RIF (Risk Influencing Factors) and their risk indicators that support risk level assessment, where indicators are used to determinate RIF states; (iii) risk indicators are obtained from monitoring; (iv) it is presented as real-time risk assessment that provides a basis for dynamic adjustments of inspection and maintenance "continuously monitor risk picture changes and support decision makers in daily operations". Chemweno et al. [4] analyze the importance of adopting DRA approaches for advanced maintenance planning decision making. These authors propose a method to introduce the dynamic approach in the failure mode and effects analysis (FMEA). In this sense it is important to highlight that FMEA/FMECA analysis is a main part of RCM analysis [26, 27], and that it is presented as one of most powerful for risk assessment by ISO 31010 [28, 29].

All these references treat DRA from the view of dynamic modification once risks have been identified, i.e., Risk Analysis (RAn) phase of RA schema of ISO 31000. However, dynamic risk changes can emerge from one or the combination of three possibilities:

1. New risks identification. Changes in the system or in its operational context (real or potential events, activities, and circumstances) appear, promoting a risk identification actualization. This drives to modifications in the risk lists and promotes a RA revision. A clear example is the treatment of "significant changes" in the EU 402/2013, common safety method for risk evaluation and assessment for railway infrastructure [30].
2. Modification in the probability of different events, activities and circumstances included in the risk modelling. Even without new risks identification, the system

could accumulate new data (new failure events modifying probability calculations) or new knowledge could emerge (new analysis results verification, organization expertise rise or new skills provided by technological development are able to be included here). This is the most extended approach in literature when DRA is addressed.

3. Modification of the consequences and/or the evaluation method of the magnitude of consequences. Not only new consequences, not considered before, can emerge, but also the organization or business priorities can change over time, and the re-evaluation of risk becomes a must [31].

In the following Sections of this Chapter, the assessment of a failure mode consequences or effects is assumed to be static [31], while the degradation process increases the risk level by increasing failure mode probability. On the other hand, maintenance tasks affect the degradation evolution, including the chance of recovering the health of the components with regard to that failure mode (from 0 to 100% depending on cases) [29].

8.5 A Failure Mode Dra Method

In this Section a method for the failure mode risk level evaluation and control is proposed, that can be used later in Chap. 9 for activities scheduling purposes. The method is based on a formal interpretation of the level of risk of a failure mode through simple business rules. The method is based on different types of events and states related to a failure mode. These key concepts for correct interpretation of the method are presented in Table 8.2.

Four failure mode risk levels or states are proposed: low, medium, high and failure; and two different types of events: monitoring and preventive maintenance events. It is considered that both monitoring events and PM events may lead to a change in the risk level of one or more failure modes of the asset. This is because these events trigger a new risk assessment of the affected FMs. A given event may affect different failure modes and in different ways. It is also assumed that reaching a new failure mode state triggers a maintenance action (the release of an algorithm for detection or prediction, an inspection, a replacement, etc.).

As an example, to show these concepts graphically, CBM over two failure modes is represented in Fig. 8.3. It is considered that both Monitoring Events and PM Events can promote a change in the risk level in one or more of the failure modes of the asset (FM1 and FM2 in the simple proposed case). These events trigger a risk analysis revision to assess the new levels of risk of the affected FMs. Note that the same event can affect different failure modes and differently. In the example a monitoring solution provides information that can be used as a symptom of FM1 and FM2. This is a common case when we are using a non-specific signal, such as global temperature, or global energy consumption. These kinds of variables are

Table 8.2 Key concepts for the failure mode DRA-based method

Concept	Types
Event Recordable, scheduled, or supervening moment, in which the risk level of the affected failure modes has to be re-analyzed	• *Monitoring Event, MoEv*: Events whose occurrence is programmed from the information generated from the condition monitoring process (including direct monitoring and PdA/PHM algorithms). They can be detection events, diagnostics events or prognostics events • *Preventive Maintenance Events, PMEv*: Maintenance task execution events whose occurrence can be programmed or unforeseen. They can be for example inspections in workshops, on board or any PM activity
State Qualitative level of risk at a given time. Each event causes a possible change in the level of risk	• *Fault*: State after the failure has occurred. State in immediate replacement or repair of the item is required • *High Risk*: State of operation closest to failure. Short-term activities are scheduled to reduce the level of risk • *Medium Risk*: State in which an anomaly has been detected but with some security it is possible to continue operating under normal conditions. Medium-term activities are planned to confirm the risk and analyze how it evolves • *Low Risk*: Normal operating state of the item
Failure Mode Failure modes involved that can be fully or partially managed by CBM. Monitoring solutions and maintenance tasks are applied at failure mode level	• Primary failure mode (PFM)[a] • Secondary failure mode (SFM)[a]: initiated by a PFM

[a] Terminology adopted from ISO 13381, (ISO, 2015)

usually good for detecting that something is happening but not so good for making accurate diagnostics (what FM is evolving).

In the example of Fig. 8.3, we represent a monitoring solution that provides information that can be linked with FM1 and FM2 risk level control. So, monitoring events (Event 1 and 2) can be related to one of three situations: FM1 is activated, FM2 is activated or FM1+FM2 are activated. Event 1 (first threshold over monitoring variable or PdA algorithm) is programmed to detect potential initial problems early, but maintainer interpretation tells us that risk level, actually, does not change, or more accurately, does not change enough to be needed to raise the level of risk of FM1 or FM2. In terms of maintenance planning, no new maintenance tasks or no changes of maintenance plan have to be planned from Event 1. Even so, Event 1 provides the maintainer useful information in order to pay attention to FM1 and FM2 from this moment, and also for the maintenance analyst to extract conclusions

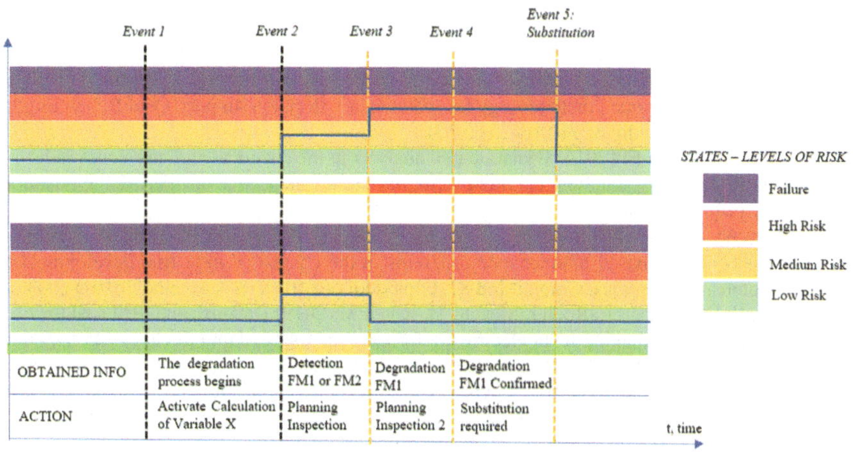

Fig. 8.3 Graphic representation of the key concepts of the event-state methodology

about the expended time by the system to get Event 2 from Event 1. With Event 2, the interpretation of the maintainer makes FM1 and FM2 reach Medium Risk level. Since it is not possible by this monitoring solution to distinguish FM1 from FM2, the maintainer introduces in the PM program an inspection, which is Event 3. Therefore Event 3 is a PM Event, concretely, an inspection task. Event 3 confirms FM1, so FM2 risk level turns back to green. At the same Event 3, the maintainer, from the inspection information report, decides to change FM1 risk level into red and to program the substitution of the component affected (Event 4).

8.6 Final Remarks

In this Chapter, a set of concepts, their interpretation and use were reviewed. How to quantify the risk levels by interpreting data and information available is a relevant engineering problem. With the use of PdA is possible to build more sophisticated tools when needed. Nevertheless, this method can be used to represent complex PdA results using a common interface. This interface can directly link maintenance decision-making related to complex assets failure modes and components.

References

1. Zio E (2018) The future of risk assessment. Reliab Eng Syst Saf 177:176–190. https://doi.org/10.1016/j.ress.2018.04.020
2. Of R, Draft T, Invited ARE, Comments WT, Any OF, Patent R et al (2009) ISO 55000 2009

3. Sola Rosique A, Crespo Marquez A (2016) Principios y marcos de referencia de la gestión de activos
4. Chemweno P, Pintelon L, De Meyer AM, Muchiri PN, Van Horenbeek A, Wakiru J (2017) A dynamic risk assessment methodology for maintenance decision support. Qual Reliab Eng Int 33:551–564. https://doi.org/10.1002/qre.2040
5. Paltrinieri N, Khan FI (2016) Dynamic risk analysis in the chemical and petroleum industry: evolution and interaction with parallel disciplines in the perspective of industrial application. Butterworth-Heinemann
6. Guillén AJ, Crespo A, Gómez JF, Sanz MD (2016) A framework for effective management of condition based maintenance programs in the context of industrial development of E-Maintenance strategies. Comput Ind 82. https://doi.org/10.1016/j.compind.2016.07.003
7. Fumagalli L, Cattaneo L, Roda I, Macchi M, Rondi M (2019) Data-driven CBM tool for risk-informed decision-making in an electric arc furnace. Int J Adv Manuf Technol 105:595–608. https://doi.org/10.1007/s00170-019-04189-w
8. Lee J, Bagheri B, Kao HA (2015) A cyber-physical systems architecture for Industry 4.0-based manufacturing systems. Manuf Lett 3:18–23. https://doi.org/10.1016/j.mfglet.2014.12.001
9. CEN-EN 13306 (2017) Maintenance-Maintenance terminology
10. Vachtsevanos G, Lewis F, Roemer M, Hess A, Wu B (2007) Intelligent fault diagnosis and prognosis for engineering systemshttps://doi.org/10.1002/9780470117842
11. Guillén AJ, Crespo A, Macchi M, Gómez J (2016) On the role of Prognostics and Health Management in advanced maintenance systems. Prod Plan Control 27:991–1004. https://doi.org/10.1080/09537287.2016.1171920
12. Zeng Z, Zio E (2018) Dynamic risk assessment based on statistical failure data and condition-monitoring degradation data. IEEE Trans Reliab 67:609–622. https://doi.org/10.1109/TR.2017.2778804
13. Villa V, Paltrinieri N, Khan F, Cozzani V (2016) Towards dynamic risk analysis: a review of the risk assessment approach and its limitations in the chemical process industry. Saf Sci 89:77–93. https://doi.org/10.1016/j.ssci.2016.06.002
14. Petchrompo S, Li H, Erguido A, Riches C, Parlikad AK (2020) A value-based approach to optimizing long-term maintenance plans for a multi-asset k-out-of-N system. Reliab Eng Syst Saf 200:106924. https://doi.org/10.1016/j.ress.2020.106924
15. Xing J, Zeng Z, Zio E (2019) A framework for dynamic risk assessment with condition monitoring data and inspection data. Reliab Eng Syst Saf 191:106552. https://doi.org/10.1016/j.ress.2019.106552
16. Aven T (2016) Risk assessment and risk management: Review of recent advances on their foundation. Eur J Oper Res 253:1–13. https://doi.org/10.1016/j.ejor.2015.12.023
17. Haugen S, Vinnem JE (2015) Perspectives on risk and the unforeseen. Reliab Eng Syst Saf. https://doi.org/10.1016/j.ress.2014.12.009
18. Aven T (2012) Foundational issues in risk assessment and risk management. Risk Anal 32:1647–1656. https://doi.org/10.1111/j.1539-6924.2012.01798.x
19. ISO 31000 (2009) International organization for standardization ISO 31000: risk management-principles and guidelines 36
20. Marquez AC, Fernandez JFG, Fernández PMG, Lopez AG (2020) Maintenance management through intelligent asset management platforms (IAMP). Emerging factors, key impact areas and data models. Energies 13. https://doi.org/10.3390/en13153762
21. Negri E, Fumagalli L, Macchi M (2017) A review of the roles of digital twin in CPS-based production systems. Procedia Manuf 11:939–948. https://doi.org/10.1016/j.promfg.2017.07.198
22. Gómez JF, Fernández PM-G, Guillén AJ, Márquez AC (2019) Risk-based criticality for network utilities asset management. IEEE Trans Netw Serv Manag 16. https://doi.org/10.1109/TNSM.2019.2903985
23. Society for Risk Analysis (SRA). Society for Risk Analysis Fundamental Principles 2018:5.
24. Khan F, Hashemi SJ, Paltrinieri N, Amyotte P, Cozzani V, Reniers G (2016) Dynamic risk management: a contemporary approach to process safety management. Curr Opin Chem Eng 14:9–17. https://doi.org/10.1016/j.coche.2016.07.006

25. Khakzad N, Khan F, Paltrinieri N (2014) On the application of near accident data to risk analysis of major accidents. Reliab Eng Syst Saf 126:116–125. https://doi.org/10.1016/j.ress.2014.01.015
26. Izquierdo J, Crespo Márquez A, Uribetxebarria J (2019) Dynamic artificial neural network-based reliability considering operational context of assets. Reliab Eng Syst Saf 188:483–93.https://doi.org/10.1016/j.ress.2019.03.054
27. Crespo Marquez A, Parra C (2012) Propuesta de un Modelo de Gestión del Mantenimiento (MGM) e Integración con el proceso de Gestión de Activos (Técnicas de Ingeniería de Mantenimiento y ... Gestión de Activos)
28. ISO (2009) ISO 31010 Risk management—risk assessment techniques
29. Erguido A, Crespo Márquez A, Castellano E, Gómez Fernández JF (2017) A dynamic opportunistic maintenance model to maximize energy-based availability while reducing the life cycle cost of wind farms. Renew Energy 114:843–56https://doi.org/10.1016/j.renene.2017.07.017
30. On the common safety method for risk evaluation and assessment and repealing Regulation 2013;EU No 402/
31. Crespo A, Sola A, Moreu P, Gómez JF, de la Fuente A, Guillén A et al (2018) Criticality Analysis for improving maintenance, felling and pruning cycles in power lines. IFAC-PapersOnLine 51:211–216. https://doi.org/10.1016/j.ifacol.2018.08.262

27. [illegible]
28. [illegible]

Chapter 9
The Dynamic Scheduling of Maintenance

9.1 Introduction

The replacement of many predetermined maintenance tasks with CBM tasks represents a change in the scheduling on maintenance interventions. Asset's condition inspections are growing, frequently becoming on-line inspections, and the maintenance service is accomplished based on the status of the asset. The fact is that many maintenance interventions will be planned and scheduled dynamically. Thus, understanding how assets conditions deteriorate, in an accurate manner, will help importantly to define (dynamically) the best moment for intervention [1].

Engineering systems with high capitalization in assets may be impacted by the way maintenance interventions are scheduled. Because of that, different researchers have been working in this area, and the intention of this Chapter is to review the work done until now, especially in two very important sectors where most of the contributions related to the introduction of new 4.0 technologies, can be found: manufacturing [2–6] and transportation [7–10].

The manufacturing dynamic scheduling papers reviewed have brought to light very interesting gaps and opportunities. Some of these refer to, for instance, the use of Digital Twin modelling to drive the dynamic scheduling of a manufacturing environment [2]. In this work multi-dimensional Digital Twin model (composed by sensor and simulated data) is used to simulate the availability of the assets and reschedule maintenance accordingly. It focuses on the operational factors of the assets (mainly availability), but from the maintenance perspective it does not rely on any dynamic condition that varies depending on the anomaly detected. The anomaly is detected but if it does not provide information about the failure mode or severity of the failure. Another interesting approach can be found in [3], where a semi-dynamic maintenance scheduling model is used to carry out Opportunistic Maintenance with condition indexes to schedule maintenance dynamically. This paper presents a time–based maintenance window combined with the failure rate of the assets to reduce maintenance costs and improve availability. As in the previous case, the root case that generates the failure rate (condition) of the asset is not defined, and the failure

A. Crespo Márquez, *Digital Maintenance Management*, Springer Series in Reliability Engineering, https://doi.org/10.1007/978-3-030-97660-6_9

rate is built upon historical and operation data using proportional hazards models. Essentially, it highlights the impact of dynamically grouping maintenance activities and the evident impact of it on the general production schedule.

In various other works such as [4] the model to optimize, dynamically, production scheduling, determines statistically the optimal preventive age of the asset. Then, an off-line scheduling optimization model generates an initial schedule, and the last stage recalculates the schedule based on unplanned disruptions or behavior from the production floor (event driven). Even this work does not provide a dynamic condition monitoring solution and it is just reactive instead of predictive; it does provide a model to redistribute production and generate an optimal schedule depending on events and makes evident that dynamically scheduling of production without considering dynamic evolution of condition of the assets is an incomplete solution. The same conclusion is reached by [5], highlighting the complexity that the asset management brings to the scheduling problem as one of the main constraints for it. Finally, the piece of research by [6] provides a complete and valuable architecture approach of the Digital Twin (DT) concept to assess the uncertainties of assets condition, processing times, and maintenance times upon prediction of failure, while scheduling production. This last reviewed work evidences the importance of the failure detection, classification and prognosis to further schedule maintenance that will (both predictive and dynamically) constitute a constraint for production scheduling.

Transportation (by train, ships, aircraft, etc.), is another sector where assets are more capitalized, critical and have a wider impact on operations and business (not to mention safety). In transports some of the assets compose the fleet (for instance the trains) and others are part of the infrastructure (for example the railway network). The fleet adds the *location* and *assignation* problem to the existing reviewed scheduling problems in manufacturing systems.

In some works, such as in [7], the authors propose a dynamic algorithm to maximize the availability of rail network depending on fleet scheduling and infrastructure maintenance. This work considers static demand, and standard fixed maintenance operations. It highlights the difficulties generated by combining maintenance scheduling and resources allocation of the fleet. On the same line, in [8] the problem is to find a flexible scheduling of the fleet depending on demand and condition of the assets. However, the condition of the assets is only based on the mileage and hours of operation, without field data, so there are numerous factors that can affect the assets degradation that are not considered. The paper focuses on the operative problem that emerges to manage the condition of the fleet together with the allocation of resources. The complexity of this problem is also explored by [9]. They add the importance of not only maintenance and assignation, but also the dependence on crew, operation demand, number of assets in the fleet, and the capabilities to carry out maintenance on an aircraft fleet. A similar direction can be appreciated in [10] where a genetic algorithm is applied dynamically to schedule the fleet (only based on operating conditions), highlighting as further work for a complete solution the dynamic allocation of maintenance intervention combining scheduled with the unexpected events and degradation.

Finally, [11] is a paper where the focus is on the detailed detection and classification of anomalies, but it is not scaled to the fleet level, making it evident that the challenge is to combine this level of detailed modelling from the maintenance perspective, combined with other dimensions of the problem of managing the fleet efficiently.

After the review in previous paragraphs, some clear conclusions arise. The dynamic scheduling of a single asset maintenance, based on its health, is a problem that has been studied numerous times and countless solutions are proposed for different scenarios. However, when the problem scales to the fleet or multi-unit level, different aspects of the problem are still very challenging. There is a clear opportunity to research in fleet scheduling, considering present operations and health conditions of the asset, to the required level of detail, identifying what failure mode causes asset's degradation, and how it evolves. This need of scaling the level of detail of the analysis, lowering in the asset indenture level of interest for the analyst, can be a critical constraint for advanced dynamic maintenance 4.0 sort of scheduling. If not accomplished, this may result in a limitation to schedule the allocation of the resources existing in the system properly.

9.2 CBM Applied to Multi-component Systems

In the previous section, dynamic maintenance scheduling within a maintenance 4.0 environment is reviewed, and the problem of managing the health of numerous assets simultaneously comes up immediately. This also generates another question: When referring to managing "the health of an asset", does it refer to a unique degradation of a component, a system or a set of systems? To answer this question, and others related to the possible interaction between degradations of components, this section reviews the application of CBM solutions for multi-component systems.

While CBM is not a novel solution itself, CBM in the PHM framework, involving prognostics in maintenance decision making for complex systems, represents novelty [12]. The application of CBM to multi-component systems grouping maintenance actions and considering the interrelations between degradations is an area to explore. For this purpose, according to [12] three main interactions must be considered: economic, structural, and stochastic. These different types of dependencies guide to consider that a CBM framework for multi-component systems should leverage multi-source dynamic information for better informed decisions, helping to predict the future behavior of the system and minimizing maintenance costs. In [13] the authors present a close solution focusing mainly on economic dependence using a multi-stage integer program. Closely related, [14] present CBM strategies for multi-unit systems studying independent degradation of components and underlining the importance of defining the possible degradation between components and the homogeneity or heterogeneity in degradation of different components in a system. So, the interdependencies between components in a complex system, together with the relations between degradations are critical factors to consider while developing a solution.

The mentioned relation between states of degradation is the core element of a considerable group of papers trying to address this challenge. In [15] an approach to optimize the maintenance of multi-component systems is presented, where the state of certain components could affect the rate of degradation of other components, defining "degradation interactions" as the interaction between states of degradation and rates of degradation. This modelling of interactions represents a great novelty. Then [16] studies these interactions considering a combination of continuous degradation that can be accelerated or have discrete changes to more degraded states depending on shocks or non-monitored events. The evident difficulties to model degradation interactions also imply that the prognosis of RUL of a system of components is complex [17]. It is relevant to mention that the evolution of the failure in one component can affect the RUL of the other and [17] present the importance of defining the different degradation modes, the moments of detection, and a clear identification of each failure mode will allow the clarification of the predictions of the system RUL. Another interesting approach can be found in [18], modelling a multi-component CBM framework that incorporates economic dependency, stochastic dependency, and imperfect maintenance (considering that some maintenance interventions do not return the degradation state to 0).

Based on the aforementioned, it can be concluded, that properly defining the object of study (asset, component, etc.) of a CBM program is essential. Furthermore, it is even more relevant when it scales from a component to a multi component system and can be considered analogue for assets. Defining the failure modes, type of evolution of degradation, prognosis of the RUL and its evolution, different type of maintenance interventions; is a critical part that then must add the dependencies related to a complex system. Due to the uncertainties the previously mentioned concept of "opportunity" is relevant to program maintenance interventions too. To a large extent, the different units that compose a system are not isolated and the way their performance and degradations behave has its effect on the system, generating the necessity of proving which level of representation is optimal to manage systems maintenance more effectively.

9.3 Fleet Maintenance Scheduling

As previously reviewed and mentioned, when dynamically managing the maintenance of multiple assets, other problem dimensions and restrictions, that are not present in a single asset maintenance scheduling problem, arise. This Section reviews contributions related to the fleet management problem focusing on maintenance scheduling. As defined by [21], a fleet is considered a set of multiple assets that fulfil the same function together for a larger service requirement. These assets can be components of a transportation fleet or for example different assets present in a production plant. Specifically, [19] define fleet maintenance management as the process of scheduling and allocating resources to the maintenance activities associated to multiple assets, highlighting the importance of maintenance scheduling together with service scheduling, opportunistic maintenance policies, effects of obsolescence and new technologies, repairing vs replacement actions, and maintenance resources allocation (capacity restrictions).

An interesting first approach to mention [20], proposes a mathematical model to systematically schedule maintenance actions for the multiple assets in the fleet presenting maintenance costs saving and prolongation of the asset's average life. This is a novel approach to model multi asset systems with complex performance and resource dependencies, balancing cost and risks as constraints. The authors clarified the different maintenance intervention levels from component, multi-component, and their dependencies, to the extension to the asset and multiple-assets level. The effect of performance dependencies from one asset to other in the fleet or the overall system, are highlighted together with the resource dependence (financial, human, and physical resources) that are shared among the assets in a fleet. However, the data to validate is not real-time condition data, so the dynamic condition-based optimization model is not fully implemented to manage the maintenance based on data from the field, the degradations are considered homogeneous, and the assets work in parallel with the failure of one meaning the stoppage of the system. Following this article, [21] also presents a maintenance strategy optimization approach for systems with parallel assets under continuous degradation. It is presented using a Bayesian framework to update unknown parameters in a Wiener degradation model. They place emphasis on the importance of the degradation models, and the lack of studies discussing maintenance policies based on performance. This emphasizes that the performance of all the assets in the fleet over time is of great interest to maintenance decision-makers. Apart from the asset management perspective, other operation and business dimensions also must be considered to provide a more integral solution, considering the opportunistic maintenance strategies depending not only on degradation but also in resource constraints from these dimensions.

The reviewed articles show the importance of Prognostics and Health Management when managing a fleet of assets. It is reduced not only to CBM but extended to operation and performance that conditions in the way the system works. In [22] the authors apply a process of managing the health of the assets for system performance impact assessment. This process highlights the importance of managing both maintenance and performance together. However, the scalability to fleet level is complicated due to the large amounts of data. So, the integral solution will then be to start from anomaly detection, diagnosis, and prognosis at the asset level (the CBM data and the importance of the failure mode), and then manage this information from all the assets together with their interdependencies and operational and resource constraints (convert prognosis results into actionable insights that optimize operational plans). Both [21, 22] represent complementary parts, two faces of the same coin. It must also be mentioned that [22] PHM methods for fleets are more useful when applied to homogeneous fleets of assets, however, even when the assets are equal, they may still present variations depending on their ages and utilization histories, and the ambient conditions [29] in which they have been working. It is also highlighted in the importance of Cyber Physical Systems (CPS) and Digital Twins (DT) as tools that can further manage fleet maintenance and operation needs, in a integrated manner.

Another interesting point is presented by [23] proposing a PHM framework for fleets of geographically distributed assets. The paper is focused on maintenance management, especially on the restriction that maintenance resources availability constitutes for managing the fleet more effectively. The other insight of the paper, is the fact that it evaluates the potential failure impact on fleet production/service loss, linking condition prognosis with operation/service. On a similar line [26] proposes a solution to keep decision makers updated with a relevant synthesis of information from both global health of the fleet and the status of maintenance efforts. This comes as a combination from monitoring and modelling of the fleet, as a knowledge scheme for PHM management and CBM in the PHM framework. The combination of both is essential to solve scheduling of maintenance as a constraint for fleet operation scheduling. Although it is conceptualized, it cannot be found to be an end-to-end solution, from data from the field and maintenance, for higher levels of fleet management.

Taking into account everything mentioned above, the multi-dimensional character of the fleet management field is clear. Fleet managers cannot ignore the fact that their decisions in terms of maintenance will impact on both company service and financial performance [24]. Although it can be stated that maintenance scheduling decisions must be based on existing data, the complexity of multiple-assets maintenance makes this task very difficult, and frequently cannot be solved by using past data to statistically plan the actions to do [24]. Due to this, there are some solutions that focus on updating real-time fleet component prognostics, associating dynamic maintenance actions to them [25].

9.4 Case Study. Scheduling Train Components' CBM

9.4.1 Introduction to the Problem

This case study is about the dynamic scheduling of maintenance activities, for a fleet of train critical components, within a maintenance 4.0 scenario. This case study tries to show only a part of this problem's complexity, in real world more considerations and constraints must be taken into account.

Maintenance 4.0 scenario means that, for each component considered, a set of algorithms will be in place, providing anomalies detection, diagnosis, and prognosis (RUL), for the existing components' failure modes. In this exercise, a continuous time system dynamics simulation model is used to deal with the problem of scheduling. Assuming that workshop capacity will be constrained, the model will find out different possibilities to balance capacity utilization, through opportunistic maintenance based on current number of components at risk.

In this exercise the time horizon for the analysis will be set to P time periods, where a period t, is defined as the time between two consecutives train safety inspections (SI), established according to existing certified maintenance programs meeting current train regulations for safety.

When the train arrives at the workshop for a regular SI, condition-based maintenance activities can be carried out on the selected components. These activities will consume workshop resources and capacity constraints may appear. The problem is to determine what are the components under the CBM program, that must be maintained in each period t.

To simplify problem formulation, it is assumed only one failure mode per component. A detected anomaly in component i, of system j, of a train k at period t, would be an input to the problem ($a(i, j, k)_t$), beside the RUL of that failure mode ($RUL(i, j, k)_t$). Notice that $a(i, j, k)_t$ is defined as a Boolean variable, that takes value 1, when an anomaly is detected for a component during a certain period t, and 0 otherwise. $RUL(i, j, k)_t$ for this specific case study, assumes that "useful life" means life of the component without the intervention of the Train Control Monitoring System (TCMS), which is a sort of safety back-up. The TCMS intervention implies, generally, penalties in train operation contracts due to stops or speed reductions to control potential failures presence.

9.4.2 Building the Continuous Time Simulation Model

In this model stock, flow and some auxiliary variables will be used. Stock variables are defined at a given time t and will remain constant unless flow variables modify them. Flow and auxiliary variables are defined for a period t. A period t is defined as the period of time between times $t - 1$ and t.

Let us assume, for instance, that a component's RUL is set to 30.000 kms while train SIs are scheduled every 10.000 km (common in many situations where SIs are released on a weekly basis). This would mean that maintenance managers will have three opportunities to do the CBM activity during a train stop for a SI service, once the anomaly is detected, and before the component's RUL expires.

In this case, for modelling purposes, three different levels of risk of a component will be modelled, once the component presents an anomaly: low, medium, and high risk. To model components under these circumstances the following stock variables are considered:

$x(j, k)_t$: Number of components with low risk in system j, train k at time t

$y(j, k)_t$: Number of components with medium risk in system j, train k at time t

$z(j, k)_t$: Number of components with high risk in system j, train k at time t

As described in Fig. 9.1, transition from no-risk or optimal condition level to low-risk levels are given by the number of positives (anomalies detected) found in a period t, see Eq. 9.1. This is considered input data to the model, and the variable will be randomly generated according to de average age of the fleet of components.

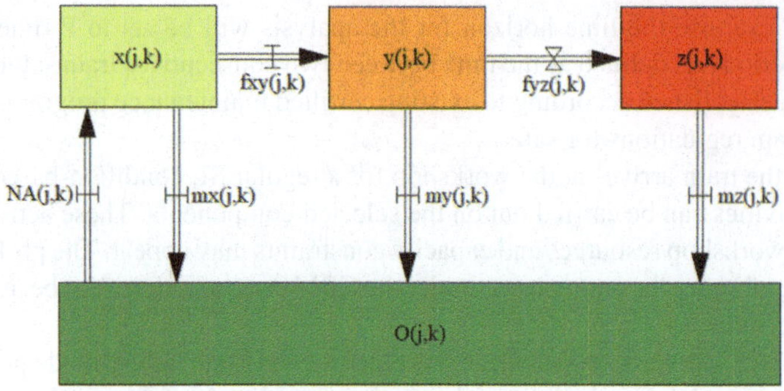

Fig. 9.1 Basic stock and flow component variables considered

$$NA(j,k)_t = \sum_{i=1}^{i=n} a(i,j,k)_t \quad (9.1)$$

The balance equations of the variables defining the components at risk (stock variables) are:

$$x(j,k)_t = \begin{cases} x(j,k)_{t-1} + NA(j,k)_t - mx(j,k)_t - fxy(j,k)_t, \; fort > 0 \\ x_0, \; fort = 0 \end{cases} \quad (9.2)$$

$$y(j,k)_t = \begin{cases} y(j,k)_{t-1} + fxy(j,k)_t - my(j,k)_t - fyz(j,k)_t, \; fort > 0 \\ y_0, \; fort = 0 \end{cases} \quad (9.3)$$

$$z(j,k)_t = \begin{cases} z(j,k)_{t-1} + fyz(j,k)_t - mz(j,k)_t - fxy(j,k)_t, \; fort > 0 \\ z_0, \; fort = 0 \end{cases} \quad (9.4)$$

where $mx(j,k)_t, my(j,k)_t \& mz(j,k)_t$ are the components at low, medium, and high risk to be maintained in period t (flow variables), and $fxy(j,k)_t \& \; fyz(j,k)_t$ represent the flow of components from a given risk level to the higher one.

Assumption 1 Components not maintained (replaced in the case of bearings), at a given risk, during one period, will move to the higher risk level in the subsequent period, i.e.:

$$fxy(j,k)_t = x(j,k)_{t-1}$$

$$fyz(j,k)_t = y(j,k)_{t-1}$$

Let us denote $O(j,k)_t$ the number of components that are running in optimal condition in system j of train k at time t. Then:

$$O(j,k)_t = \begin{cases} O(j,k)_{t-1} - NA(j,k)_t + mx(j,k)_t + my(j,k)_t + mz(j,k)_t, \ for\ t > 0 \\ O_0, \ for\ t = 0 \end{cases} \tag{9.5}$$

Obviously:

$$O(j,k)_t + x(j,k)_t + y(j,k)_t + z(j,k)_t = n, \forall t \tag{9.6}$$

Assumption 2 Workshop capacity, now denoted $c(i, j)$ will be balanced to service all components and have them running before their RUL expiration. $c(i, j)$ can be defined as the maximum number of components of system j that can be maintained during a train k SI stop. Thus, the intention is that.

$$mx(j,k)_t + my(j,k)_t + mz(j,k)_t \leq c(j,k), \forall t, j, k \tag{9.7}$$

Due to the stochastic nature of $NA(j,k)_t$, besides the fact that many components could increase in risk level, with a higher probability, at certain number of running hours or kilometres, there must be a control of anomalies and capacity, since during a certain period it could happen that.

$$NA(j,k)_t > c(j,k)$$

Let us denote $Ex(j,k)_t$ as the difference between anomalies detected and existing maintenance capacity in a SI. That is the excess of component experiencing anomalies compared to the ones that the maintenance depot could service, with the existing capacity, in one SI:

$$Ex(j,k)_t = Max(NA(j,k)_t - c(j,k), 0) \tag{9.8}$$

In case that $Ex(j,k)_t > 0$, this will force the maintenance organization to act, and to replace components before they reach their highest admissible risk level. Otherwise, capacity could not be enough to service all the components. At the same time, the idea is that lower risk components could be replaced as few times as possible to lengthen their useful lives.

To control maintenance activities over components with different risk levels, the existing accumulated excess of anomalies registered (now denoted $AEx(j,k)_t$) must be formulated and controlled.

$$AEx(j,k)_t = \begin{cases} AEx(j,k)_{t-1} + Ex(j,k)_t - LEx(j,k)_t, \ for\ t > 0 \\ AEx_0, \ for\ t = 0 \end{cases} \tag{9.9}$$

With $LEx(j,k)_t = my(j,k)_t + mx(j,k)_t$, See Fig. 9.2.

Assumption 3 All components reaching the highest risk admissible level in a period t will be serviced in that period, i.e.:

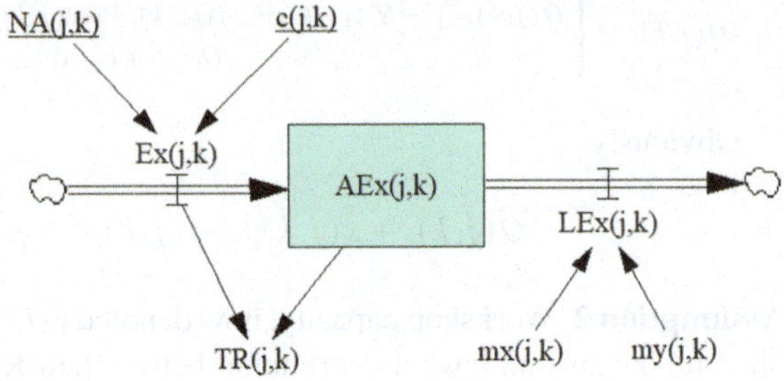

Fig. 9.2 Accumulated excess capacity modelling and control

$$mz(j,k)_t = fyz(j,k)_t \Rightarrow z(j,k)_t = 0, \forall t$$

Then notice that the only way to lower this accumulated excess of anomalies would be to increase capacity, or to replace components with lower risk levels $(my(j,k)_t \& mx(j,k)_t)$.

The total components to maintain (to replace in the case of bearings) with low or medium risk levels, during a given period t, can be determined as follows (see also Fig. 9.2):

$$TR(j,k)_t = AEx(j,k)_{t-1} + Ex(j,k)_t \tag{9.10}$$

Then, how to distribute these activities for the different risk levels, the solution to the problem formulated, can be calculated, as in the following Equations:

$$my(j,k)_t = Min(fxj(j,k)_t, TR(j,k)_t, c(j,k) - mz(j,k)_t) \tag{9.11}$$

With $c(j,k) - mz(j,k)_t \geq 0$

$$mx(j,k)_t = Min(TR(j,k)_t - my(j,k)_t, c(j,k) - mz(j,k)_t - my(j,k)_t) \tag{9.12}$$

With $c(j,k) - mz(j,k)_t - my(j,k)_t \geq 0$.

This model formulation is rather simple and does not take into consideration, so far, relationships among components generating opportunistic maintenance options. Nor the fact that different workshops could have different capacity. It is just built to see how workshop capacity may impact the scheduling of maintenance operations when CBM programs are in place.

9.4.3 Implementation of a Practical Model in Vensim

Vensim ® provides high rigour for writing model equations, helps to trace and to understand the importance of model existing feedback loops and supports multi-parametric optimization that may result essential for this work.

A practical version of the model to appreciate capacity problems when dealing with a component CBM activities has been built as in Figs. 9.3 and 9.4. This model is built for the analysis of bearing's CBM activities scheduling purposes.

Interactive simulations allow the user to understand the need to maintain the bearings with lower risk when depot capacity decreases (see Fig. 9.5).

Graphical results provided by the tool, despite being a very simple model, help to understand implications in the number of bearings at risk when changing workshop capacity (see Fig. 9.6).

In Figs. 9.7, 9.8 and 9.9, different results are obtained for different workshop or depot capacity options. The higher the capacity of the depot the lesser the number of bearings maintained with lower levels of risk.

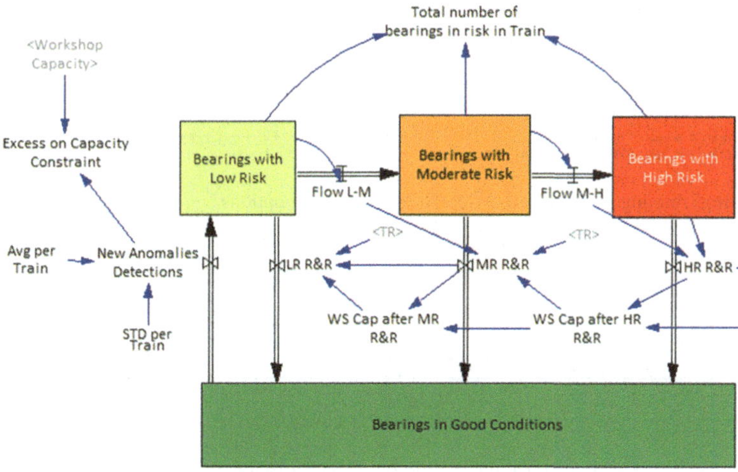

Fig. 9.3 Stock and flow diagram of the practical model

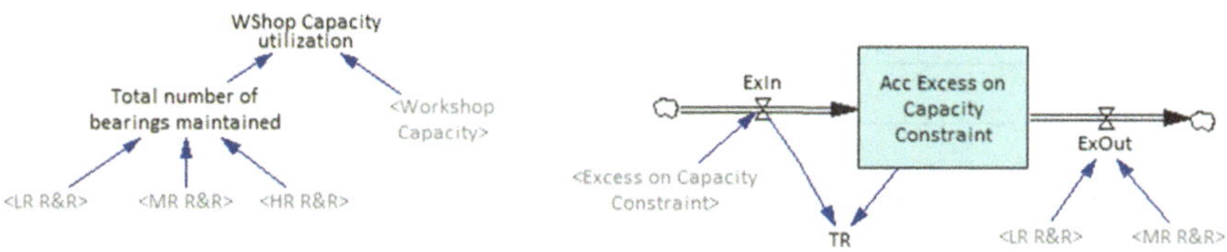

Fig. 9.4 Stock and flow diagram of the practical model (cont.)

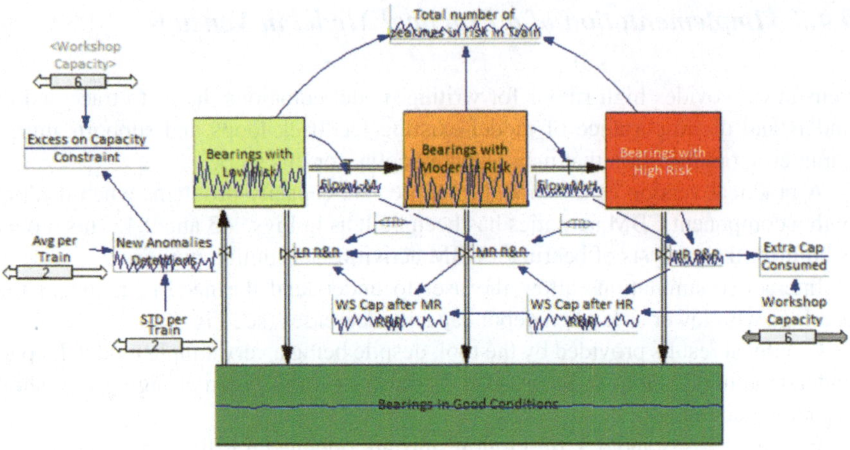

Fig. 9.5 Practical interactive simulation

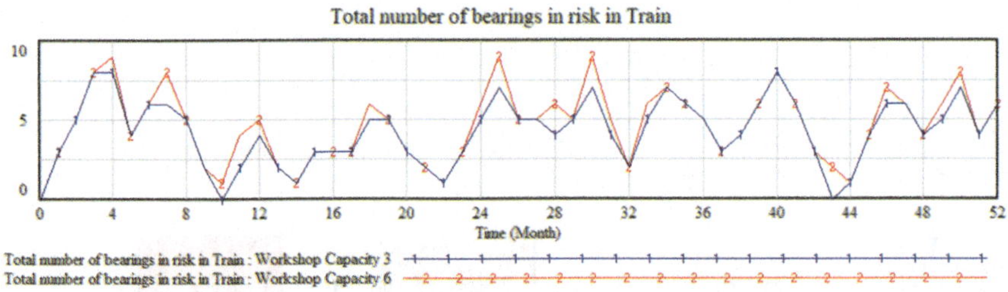

Fig. 9.6 Bearings running at risk under different depot capacity options

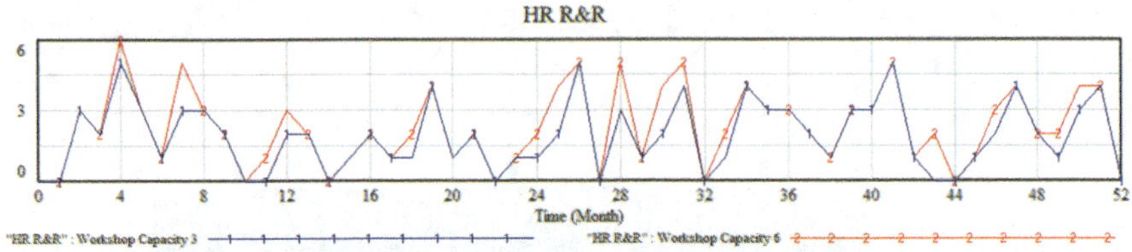

Fig. 9.7 High risk bearings maintained under different depot capacity options

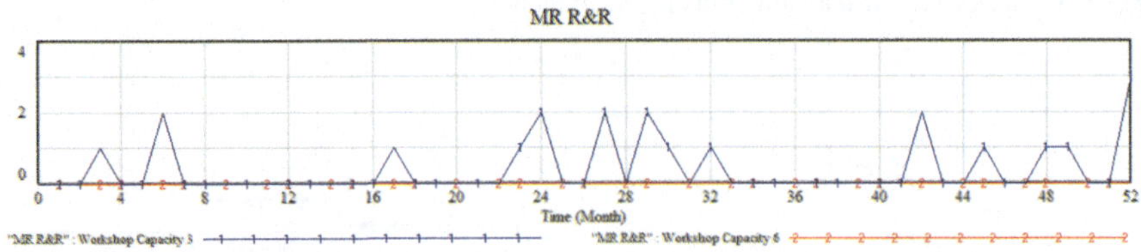

Fig. 9.8 Medium risk bearings maintained under different depot capacity options

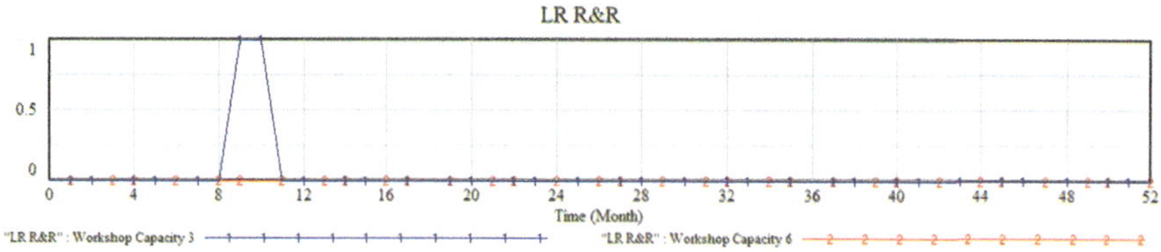

Fig. 9.9 Low risk bearings maintained under different depot capacity options

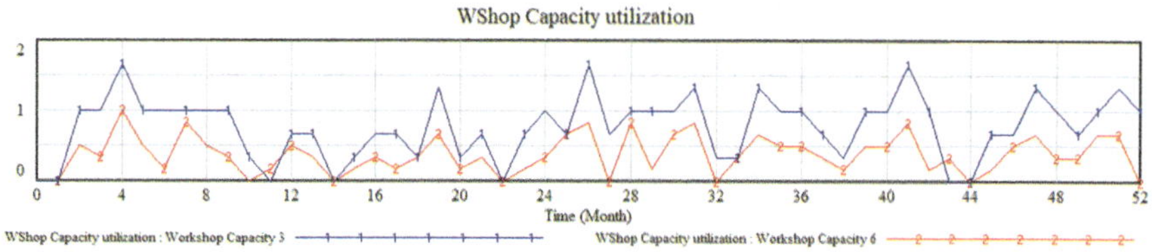

Fig. 9.10 Workshop capacity utilization with the two simulated scenarios

Figures 9.10 shows depot's capacity utilization under the two simulated scenarios. Notice that the scenario shown for Capacity = 3 bearing/period, would not be technically feasible according to bearings maintenance demand assuming maximum capacity utilization set to 1. However, the analysis could help to determine excess capacity needed in advance per workshop.

9.4.4 Conclusions of the Case Study

This exercise shows that practical simulation tools can be useful when dealing with the problem of scheduling maintenance activities. Complexity can be added when required to the case study, and maintenance opportunistic options and rules can be identified, optimized, and implemented. Given the dynamic nature of the problem, the simpler the business rules the easier their implementation at the shop floor.

9.5 Conclusions

Some important considerations should be mentioned to conclude the Chapter. These are related to the lack of a standardized strategic solution to dynamically manage the fleet, based on the condition of the assets and operational (scheduling) needs. This statement is supported by [26] that highlights the importance of the PHM system, combined with service scheduling, for a holistically optimized maintenance scheduling and management of the fleet, not only focusing on maintenance but also on operations and even other stakeholders' interest. A similar idea can be deduced from

[27], with emphasis on the difficulty of scheduling based on the RUL of asset components together with other operational factors. The importance of RUL assessment is highlighted as well as the need of defining a period for the dynamic solution recalculation (for instance, during certain period, heterogeneous degradations of different asset components in the fleet could generate different optimal decision horizons [30, 31]).

Related problems such as spare parts and other maintenance resources management, as well as the different capabilities for maintenance depending on the geographical location must be analyzed [28, 32]. These are aspects that must also be considered when developing and implementing CBM plans [29]. Also notice that fleets are often managed with a RUL index for each asset (which is defined by the most critical component), but assets could be broken down in subsystems with their components and RULs, with different criticalities. This will normally offer room for the consideration of opportunistic maintenance together with PHM in a dynamic way.

Finally, regardless of the fact that there is an uncertain duration of maintenance interventions, all these challenges must be addressed fulfilling existing policies and safety regulations [33].

References

1. Bin L, Liang Z, Parlikad AK, Xie M, Kuo W (2017) Condition-based maintenance for systems with aging and cumulative damage based on proportional hazards model. Reliab Eng Syst Saf 168:200–209. ISSN 0951-8320. https://doi.org/10.1016/j.ress.2017.04.010
2. Zhang M, Tao F, Nee AYC (2021) Digital twin enhanced dynamic job-shop scheduling. J Manuf Syst Part B 58:146–156. ISSN 0278-6125. https://doi.org/10.1016/j.jmsy.2020.04.008
3. Zhou X, Mengqi Y (2020) Semi-dynamic maintenance scheduling for multi-station series systems in multi-specification and small-batch production. Reliab Eng Syst Saf 195:106753. ISSN 0951–8320. https://doi.org/10.1016/j.ress.2019.106753
4. Ghaleb M, Taghipour S, Zolfagharinia H (2020) Real-time optimization of maintenance and production scheduling for an industry 4.0-based manufacturing system. In: 2020 annual reliability and maintainability symposium (RAMS), pp 1–8. https://doi.org/10.1109/RAMS48030.2020.9153721
5. Rossit DA, Tohmé F, Frutos M (2019) A data-driven scheduling approach to smart manufacturing. J Ind Inf Integr 15:69–79. ISSN 2452-414X. https://doi.org/10.1016/j.jii.2019.04.003
6. Negri E, Pandhare V, Cattaneo L et al (2021) Field-synchronized Digital Twin framework for production scheduling with uncertainty. J Intell Manuf 32:1207–1228. https://doi.org/10.1007/s10845-020-01685-9
7. Zhang C, Gao Y, Yang L, Gao Z, Qi J (2020) Joint optimization of train scheduling and maintenance planning in a railway network: a heuristic algorithm using Lagrangian relaxation. Transp Res Part B Methodol 134:64–92. ISSN 0191-2615. https://doi.org/10.1016/j.trb.2020.02.008
8. Agostino M, Caballini C, Chiara BD, La Scala PG (2020) Compliance of maintenance and operational needs for trains: a simulation analysis to evaluate the impact of a flexible scheduling on local transport by rail. In: 2020 IEEE 23rd international conference on intelligent transportation systems (ITSC), pp 1–6. https://doi.org/10.1109/ITSC45102.2020.9294379

9. El Moudani W, Mora-Camino F (2000) A dynamic approach for aircraft assignment and maintenance scheduling by airlines. J Air Transp Manag 6(4):233–237. ISSN 0969-6997. https://doi.org/10.1016/S0969-6997(00)00011-9
10. Sanhueza C, Mendes A, Jackson M, Clement R (2020) An efficient genetic algorithm for the train scheduling problem with fleet management.In: 2020 IEEE congress on evolutionary computation (CEC), pp 1–8. https://doi.org/10.1109/CEC48606.2020.9185779
11. Zaccaria, V, Stenfelt, M, Aslanidou, I, Kyprianidis, KG (2018)Fleet monitoring and diagnostics framework based on digital twin of aero-engines. In: Proceedings of the ASME turbo expo 2018: turbomachinery technical conference and exposition, volume 6: Ceramics; Controls, Diagnostics, and Instrumentation; Education; Manufacturing Materials and Metallurgy. Oslo, Norway. 11–15 Jun 2018. V006T05A021. ASME. https://doi.org/10.1115/GT2018-76414
12. Shi Y, Zhu W, Xiang Y, Feng Q (2020) Condition-based maintenance optimization for multi-component systems subject to a system reliability requirement. Reliab Eng Syst Saf 202:107042. ISSN 0951-8320. https://doi.org/10.1016/j.ress.2020.107042
13. Zhu Z, Xiang Y (2021) Condition-based maintenance for multi-component systems: modeling, structural properties, and algorithms. IISE Trans 53(1):88–100. https://doi.org/10.1080/24725854.2020.1741740
14. Sun Q, Ye Z, Chen N (2018) optimal inspection and replacement policies for multi-unit systems subject to degradation. IEEE Trans Reliab 67(1):401–413. https://doi.org/10.1109/TR.2017.2778283
15. Rasmekomen N, Parlikad AK (2016) Condition-based maintenance of multicomponent systems with degradation state-rate interactions. Reliab Eng Syst Saf 148:1–10. ISSN 0951-8320. https://doi.org/10.1016/j.ress.2015.11.010
16. Shen J, Elwany A, Cui L (2018) Reliability analysis for multi-component systems with degradation interaction and categorized shocks. Appl Math Model 56:487–500. ISSN 0307-904X. https://doi.org/10.1016/j.apm.2017.12.001
17. Prakash O, Samantaray AK, Bhattacharyya R, Ghoshal SK (2018) Adaptive prognosis for a multi-component dynamical system of unknown degradation modes. IFACPapersOnLine 51(24):184–191. ISSN 2405-8963. https://doi.org/10.1016/j.ifacol.2018.09.576
18. Xu J, Liang Z, Li Y-F, Wang K(2021) Generalized condition-based maintenance optimization for multi-component systems considering stochastic dependency and imperfect maintenance. Reliab Engi Syst Saf
19. Cassady CR, Murdock WP, Nachlas JA, Pohl EA (1998) Comprehensive fleet maintenance management. In: SMC'98 conference proceedings. 1998 IEEE international conference on systems, man, and cybernetics (Cat. No.98CH36218), vol 5, pp 4665–4669. https://doi.org/10.1109/ICSMC.1998.727588
20. Petchrompo S, Li H, Erguido A, Riches C, Parlikad AK (2020) A value-based approach to optimizing long-term maintenance plans for a multi-asset k-out-of-N system. Reliab Eng Syst Saf 200:106924. ISSN 0951-8320. https://doi.org/10.1016/j.ress.2020.106924
21. Zhao X, Liang Z, Parlikad AK, Xie M (2021) Performance-oriented risk evaluation and maintenance for multi-asset systems: a Bayesian perspective. IISE Trans. https://doi.org/10.1080/24725854.2020.1869871
22. Liu Z (2018) Cyber-physical system augmented PHM for fleet based systems, Doctoral Thesis, University of Cincinnati
23. Jin C et al (2015) A comprehensive framework of factory-to-factory dynamic fleet-level prognostics and operation management for geographically distributed assets. In: IEEE international conference on automation science and engineering (CASE), pp 225–230. https://doi.org/10.1109/CoASE.2015.7294066
24. Bivona E, Battista G (2005) Evaluating fleet and maintenance management strategies through system dynamics model in a city bus company
25. Xia T, Dong Y, Pan E, Zheng M, Wang H, Xi L (2021) Fleet-level opportunistic maintenance for large-scale wind farms integrating real-time prognostic updating. Renew Energy 163:1444–1454. ISSN 0960-1481. https://doi.org/10.1016/j.renene.2020.08.072

26. Van Horenbeek A, Scarf PA, Cavalcante CAV, Pintelon L (2013) The effect of maintenance quality on spare parts inventory for a fleet of assets. IEEE Trans Reliab 62:596–607. https://doi.org/10.1109/TR.2013.2270409
27. Crespo Márquez A, Moreu De Leõn P, Sola Rosique A, Gõmez Fernández JF. Criticality analysis for maintenance purposes: a study for complex in-service engineering assets. Qual Reliab Eng Int 2016;32:519–33. https://doi.org/10.1002/qre.1769
28. Wang K, Djurdjanovic D (2018) Joint optimization of preventive maintenance, spare parts inventory and transportation options for systems of geographically distributed assets. Machines 6(4):55. https://doi.org/10.3390/machines6040055
29. Meissner R, Rahn A, Wicke K (2021) Developing prescriptive maintenance strategies in the aviation industry based on a discrete-event simulation framework for post-prognostics decision making. Reliab Eng Syst Saf 214:107812. ISSN 0951-8320. https://doi.org/10.1016/j.ress.2021.107812
30. Bougacha O, Varnier C, Zerhouni N, Dersin P (2021) Impact of decision horizon on post-prognostics maintenance and missions scheduling: a railways case study. Int J Rail Transp. https://doi.org/10.1080/23248378.2021.1940329
31. Mira L, Andrade AR, Gomes MC (2020) Maintenance scheduling within rolling stock planning in railway operations under uncertain maintenance durations. J Rail Transp Plann Manag 14:100177. ISSN 2210-9706. https://doi.org/10.1016/j.jrtpm.2020.100177
32. Wang Y, Limmer S, Van Nguyen D, Olhofer M, Bäck T, Emmerich M (2021) Optimizing the maintenance schedule for a vehicle fleet: a simulationbased case study. Eng Optim. https://doi.org/10.1080/0305215X.2021.1919888
33. Dersin, P (2018) From single asset health monitoring to dynamic fleet maintenance. ALSTOM Digital Mobility

Part V
Emerging Data-Driven Processes for Asset Performance Management (APM)

Chapter 10
Techniques for Anomalies Detection

10.1 Introduction

Within the field of soft computing, intelligent optimization modelling techniques include various major techniques in artificial intelligence [1] pretending to generate new business data knowledge transforming sets of "raw data" into business value.

Nowadays, connections among industrial assets and integrating information systems, processes, and operative technicians [2] are the core of the next generation of industrial management. Based on the industrial Internet of Things (IoT), companies have to seek intelligent optimization modelling techniques (advanced analytics) [3] in order to optimize decision-making, business and social value. These techniques are preferred to fall inside the soft computing category, with the idea of solving real complex problems with inductive reasoning like humans, searching for probable patterns, being less precise, but adaptable to reasonable changes and being easily applicable and obtainable [4].

To be able to implement these advanced techniques requires a comprehensive process sometimes named "intelligent data analysis" (IDA) [5], which is a more extensive and non-trivial process to identify understandable patterns from data. Within this process, the main difficulty is to identify valid and correct data for the analysis [3] from the different sources in the company. Second, efforts must be developed to create analytic models that provide value by improving performance. Third, a cultural change has to be embraced for companies to facilitate the implementation of the analytical results. In addition to this, since accumulation of data is too large and complex to be processed by traditional database management tools (the definition of "big data" in the Merriam–Webster dictionary), new tools to manage big data must be taking into consideration [6].

In this Part of the book, interesting examples of the use of different predictive analytics techniques in emerging business processes will be presented. These are examples where the use of these new methods and techniques could successfully be translated into an increase in company profits.

A. Crespo Márquez, *Digital Maintenance Management*, Springer Series
in Reliability Engineering, https://doi.org/10.1007/978-3-030-97660-6_10

In this Chapter, an example of baseline predictive analytics for a process of train bearings anomalies detection is presented. The data used is collected from local sensors in the train when operating under normal operating conditions. In Chap. 11, two examples of the use of these techniques are presented where, in addition to the detection of anomalies in systems, the techniques are also supporting the process to assess the loss of energy generation and the reduction of equipment efficiency. Later in Chap. 12 a more strategic solution is supported for asset health determination with the purpose to support the process of reinvestment in assets.

In most of these cases, technology infrastructure (in terms of sensors and connection for the data to flow to the App responsible for the corresponding analysis) was preexisting, therefore company investment for this was not very relevant to the business compared to the return obtained.

10.2 Anomalies Detection in Train Bearings

This case study is concerned about trains' rolling system, which is one of the most demanding systems in modern trains. The running gear system of a train may consist of bogies, brake elements, wheels, wheel guiding elements, axle bearings and their greasing system, suspensions, and monitoring equipment (accelerometers and temperature probes). Within this system we can also consider traction elements, such as electric motors, and shunt elements.

Within the rolling system, axle bearings are critical from the point of view of service quality and availability to the customer. A bearing failure may cause a train to stop, even to enter in "useless train" status, which normally represents relevant economic losses according to the contractual conditions of the service. In addition, delays and interruptions on the railway network could be generated by the same incident too.

Concerning the axle bearings, leading manufacturers in this sector customize the optimal bearing to fit a given train model based on a wide space of considerations: radial and axial loads, track conditions, environment, and type of service; railcar and truck type; track gauge; type of cargo; among others. Although the failure rate of the elements making up these systems is very low—with most of element's incidents registered with a rate of less than $5 \cdot 10^{-3}$ failures per million kilometers (FPMK)—more than 50% of these elements are considered critical elements given the severity of their functional loss.

In the relevant literature, new contributions are showing up, developing advanced analytics-enabled framework for monitoring the deterioration of these systems, predicting their rate of deterioration, and recommending the optimal scheduling of their maintenance operations to maximize expected lifetime [7]. This type of analytics, known as Predictive Analytics (PA), scores data using predictive models, that can be grouped into regression and machine learning (ML) models, the latter being chosen in this work since the underlying relationships among variables tend to be very complex and the mathematical form of the dependencies unknown [8].

Improvements in computing and communication technologies allow automated data collection on a large scale. Prediction models can then be conveniently trained and put into operation with real-time sensors data [9].

Businesses' digital transformation fosters a new O&M paradigm: merging data collected from assets and sensors with big data analytics allows to monitor entire fleets down to individual components and plan maintenance actions only when needed [10].

Axle bearings are, in many train configurations, monitored through either temperature, vibration or both condition measurements. These two measurements really complement each other with great efficiency in installations with technically expensive bearings [11]. The vibration measurement is convenient for fatigue damage analysis (for instance in bearings with point contact or ball bearings). However, these procedures are less used for the observation of the lubrication conditions. A failure of the lubricant condition is reliably recognized by a temperature measurement. The measurement of the shock pulse reacts remarkably less sensitively than the temperature sensor in these cases. When the overheating of the bearing takes place, this may cause damage to the seating surfaces, to the rolling surfaces, to tabs and roller faces, it may also produce cage damage and shutter. So once a deterioration in the bearing is recognized, the question arises as to whether immediate replacement is necessary or if the bearing can continue in use until the next scheduled stop of the machine. The answer to this question depends on several conditions [11]:

- Loading: In case of moderate loading, deterioration progresses slowly, so that the next scheduled stop can be expected to replace the bearing. As the load increases, the deterioration spreads much faster.
- Degree of damage: Initially, the damage grows slowly, but the increase in the degree of damage will also greatly increase the speed of propagation (Fig. 10.1).

As the reader may guess, capturing altered bearing temperature behavior is not an easy task, it requires having the capability to assess what should be a normal

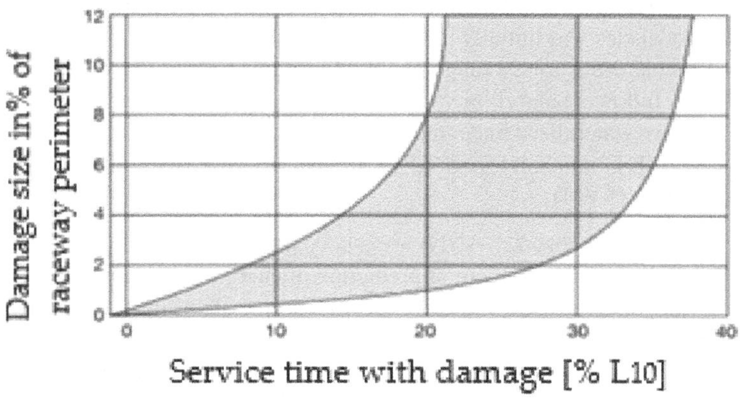

Fig. 10.1 Bearing damage size versus service time with damage

bearing temperature behavior during train operations, assuming a certain load (that depends on the weight and speed of the train, as well as on the conditions of the infrastructure), a given outside temperature and regular lubrication. For a precise study of forces acting on a wheelset, the reader is referred, for instance, to the paper of Nagumo et al. [12], helping to understand the resulting load on the rolling system bearings.

In this case study, the train configuration considered presents only bearing temperature monitoring and the idea is to develop a predictive maintenance strategy that using these sensors can capture altered bearing temperature.

Predictive analytics and big data processes and tools will be applied to design to design this strategy. The objective is to increase both preventive maintenance (PM) intervals, reliability, and availability of the trains.

The following Sections describe how a machine learning predictive model is selected, trained with different data sets (in Chap. 6, it was shown how big data processes allowed the testing and acceptance of a universal model per bearing position regardless the axle or train of the fleet, overcoming complexity that could be generated by the non-ergodicity of these assets). Interpretation rules for early failure detection based on these advanced predictive analytics are compared to those already existing rules in the train on-board control monitoring system (TCMS) ensuring train's safety. The CBM plan expected performance is assessed before its implementation to allow its further control during train operations.

10.3 Baseline Analytics Selection

Baseline analytics (Industrial Internet Consortium, 2015) to detect anomalies in train axle bearings temperature require the selection of a predictive model and its training, validating and testing for prediction.

Machine learning models were chosen in this work since the underlying relationships among variables were expected to be very complex and the mathematical form of the dependencies was initially unknown. Dynamic modelling of bearing temperature is possible using a wide range of ML algorithms options as predictive models [13–16]. The following algorithms will be introduced, because they were considered for the analysis: generalized linear models (GLM), artificial neural networks (ANN), decision trees (DT), random forest (RF), Gradient Boosted Trees (GBT) and support vector machines (SVM):

- *Artificial neural networks* (ANN) are a very popular solution to solve complex problems, such as the problem of prediction of energetic consumption in pumps. Neural networks can use a wide diversity of learning algorithms. A popular supervised model is the multilayer Perceptron trained with variations of the backward propagation algorithm (BPN). BPN is a feedback model with supervised learning [17, 18].

- *Decision trees* (DT) are tree-shaped structures representing clusters of detections that can produce class rules for a set of singular data, or as Berry and Linoff pointed out "A structure that can be used to divide a large collection of records in successively smaller sets of records applying a sequence of simple decision rules"[19]. The most used names for the models are Classification Trees or Regression Trees.
- *Random forests* (RF) are effective and intuitive models used in regression and classification problems. They are intuitive because they provide a clear path to an outcome and are based on the underlying structures of the Decision Tree. A decision tree is a model created using a series of decisions based on variable values to take one route or another. A tree depends on a random vector sampled independently. All trees' vectors in the forest are sampled using the same distribution. The generalization error for forests converges as the number of trees in the forest increases. This error of a tree-classifying forest depends on the weight of the individual trees in the forest and on the existing correlation among them. Error, strength, and correlation are estimated and used to demonstrate the response to a division with a higher number of factors. Internal metrics are also used to assess variable importance [20].
- *The gradient boosted trees* (GBT) [21] results from applying boosting methods to regression trees. It calculates a sequence of trees in which each successive tree is built from the prediction residues of the previous tree. At each step of the reinforcement algorithm, the data is divided into two samples in each divided node, determining the best partition and regression errors are calculated. Then, the next tree is adjusted to reduce the error. The driven tree improves performance of a single model, adjusting many models and combining them for prediction. To avoid overfitting, the maximum number of additive trees is set to 200. The subsample ratio for building a tree is 0.5, and the maximum number of levels is set to 10.

For the assessment of the algorithm's behavior, different performance measures were selected. These measures were divided into two groups: Error related measures [Mean Absolute Error (MAE), Root Mean Squared Error (RMSE), Relative Error (MRE), Squared Error (MSE) and Correlation] and computing time related measures (training and scoring times). In the next paragraphs a definition of each one of the measures, in that order, is presented.

- *The mean absolute error* (MAE) calculation is based on the prediction error, the prediction error ei is determined by the difference between the value predicted by the model ŷi and the actual value yi. For the prediction or residual error ($e_i = y_i - \hat{y}_i$), the MAE for a value it is given by Eq. (10.1):

$$MAE = mean(|e_i|) \tag{10.1}$$

The error of the individual data point may be positive or negative and may cancel each other out. To derive the overall prediction for the model, calculate the absolute error to aggregate all the residuals and average it. This measure is used to know the accuracy.

- In some cases, it is advantageous to penalize the individual point error with higher residues. Even though two models have the same MAE, one might have consistent error and the other might have low errors for some points and high error for other points. *Root mean square error* (RMSE) penalizes the latter. The RMSE is given by Eq. (10.2):

$$RMSE = \sqrt{mean(e^2)} \tag{10.2}$$

- The *mean relative error* (MRE) is the average of the absolute deviation of the prediction of the real value divided by the real value. This measure is used to know the quality, and the MRE is given by Eq. (10.3):

$$MRE = mean\left(\frac{(|y_i - \hat{y}_i|)}{y_i}\right) \tag{10.3}$$

- The *mean squared error* (MRE) is an estimator that measures the average value of errors squared, that is, the difference between the prediction value and the actual value. The difference occurs due to randomness or because the prediction value does not consider the information that could produce a more accurate estimate. The MSE is a risk function, corresponding to the expected value of the quadratic loss. The MSE is given by Eq. (10.4):

$$MSE = \frac{1}{n}\sum_{i=1}^{n}(y_i - \hat{y}_i)^2 \tag{10.4}$$

  Where n represents the number of points in the dataset.

- *Correlation* is a statistical measure of the relationship between two variables, it measures the dependence of one attribute on another. When the two variables are highly correlated, they both have the same rate of variation (in the same direction or in opposite directions). The correlation between the variables does not imply a causal relationship, that is, one does not necessarily cause the other to vary. This correlation is measured by the Pearson correlation coefficient (r), which measures the strength of linear dependence. Correlation between two attributes ŷi and yi and is calculated with the Eq. (10.5):

$$r_{y_i\hat{y}_i} = \frac{\sum_{i=0}^{n}(y_i - \overline{y}_i)(\hat{y}_i - \overline{\hat{y}}_i)}{\sqrt{\sum_{i=0}^{n}(y_i - \overline{y}_i)^2 \sum_{i=1}^{n}(\hat{y}_i - \overline{\hat{y}}_i)^2}} \tag{10.5}$$

- Concerning the algorithm computing time performance measures, the results of the different techniques are obtained using the Automodel option, with the educational edition of the RapidMiner Studio software Version 9.5 (copyright RapidMiner GmbH 2001–2019). Rapid miner provides *training* and *scoring* times for each one of the algorithms per 1000 rows of data in milliseconds (ms). Total time can be then considered as the sum of training and scoring time. Experiments for all algorithms were performed on a laptop with I7 processor, 8GB memory RAM and operating system 64 bits.

To test different models' performance under similar train operating conditions, several short periods of one train operation were selected to have a reasonable data extraction and to test their capabilities for temperature prediction. More precisely, data used was obtained during the following calendar time intervals: (1/02/2016—3/02/2016); (3/02/2016—5/02/2016); (5/02/2016—7/02/2016); (7/02/2016—9/02/2016); and (9/02/2016—11/02/2016). During these periods data was registered every single minute, and a total of 8,265 time events were correctly recorded, with the following filtered and cured data: Operating date and time, train speed, axel hydraulic pressure, temperature per bearing (assuming four bearings positions per axle: $T1$, $T2$, $T3\&T4$) and outside temperature (Ta).

Concerning training, 70% of the data was selected for training randomly and the remaining 30% was considered for model validation [22]. Performance obtained for the different predictive models is presented in Table 10.1.

Performance results were presented to the rolling stock company and their *Smart Maintenance Engineering and New Projects* Department determined to initially select the ANN because of the best ratio of correlation coefficient versus total time, assuming a minimum required correlation factor of 0.9600.

10.4 Anomalies Detection Model Implementation

The deployment of a strategy for each single bearing overheating failure mode requires to identify descriptors of the failure modes and interpretation rules to detect abnormal temperature behavior [23, 24]. Additionally, the benefits of the new CBM strategy implementation must be demonstrated, and for that purpose, current interpretation rules existing in the TCMS must me compared to the proposed CBM plan rules. To that end in this Section the following steps are followed:

1. Notation for the monitoring variables utilized
2. Definition of Descriptors
3. Definition of current rules implemented in the TCMS
4. Definition of rules for a possible new CBM plan
5. Benefits of the new CBM plan

Table 10.1 Performance of predictive models in the initial example

Predictive model	Root mean squared error (± SD)		Absolute error (± SD)		Relative error (%) (± SD)		Squared error (± SD)		Correlation (± SD)		Training Time (ms)	Scoring Time (ms)
Generalized linear model	2.81	0.03	2.2	0.0	0.45	0.00	7.92	0.18	0.81	0.01	53.1	5.1
Neural network	*1.62*	*0.05*	*1.2*	*0.0*	*0.30*	*0.01*	*0.47*	*0.03*	*0.97*	*0.00*	*230.2*	*57.8*
Decision tree	0.82	0.13	0.4	0.0	0.09	0.00	0.68	0.20	0.95	0.00	126.4	9.1
Random forest	2.29	0.01	1.8	0.0	0.41	0.00	5.26	0.03	0.92	0.00	465.0	20.6
Gradient boosted trees	0.68	0.02	0.5	0.0	0.18	0.00	0.46	0.03	0.99	0.00	483.0	78.6
Support vector machine	1.25	0.12	0.8	0.1	0.23	0.01	1.57	0.31	0.97	0.01	6715.3	658.8

Once benefits of the potential CBM plan are tested, the CBM plan can be precisely designed, establishing the thresholds for the interpretation rules with precision, and calculating the expected performance of the plan, to be then controlled.

The monitoring variables considered are the following:

- Ta_t: Outside Temperature at time *t*.
- TS_t: Train Speed at time *t*.
- Tij_t: Temperature of bearing position *i* in axle *j* at time *t*.

Descriptors are, beside monitoring variables, additional measures that may allow to track the evolution of the degradation. For this case study the following are taken:

- $\tilde{T}ij_t$: Predicted (ML model) temperature of bearing position *i* in axle *j* at time *t*.
- $AE(Tij_t)$: Absolute error of bearing *ij* temperature prediction at time *t*.
- $Tmax_j$ =: Maximum temperature among same axle bearings at time t

$$Tmaxj_t = Max_i(Tij_t)\forall i$$

With i = 1…n bearing positions per axle; in our case study n = 4.
- $Tmin_j$: Minimum temperature between the four bearings of the same axle

$$Tminj_t = Min_i(Tij_t)\forall i$$

- μTi_t: Average bearing temperature for all Ti bearings in the train at time t:

$$\mu Ti_t = \frac{\sum_{j=1}^{j=m} Tij_t}{m} \tag{10.6}$$

With m = number of train axles, in our case study m = 21.
- σTi_t: Standard deviation of bearing temperature for all $T1$ bearings in the train at time t:

$$\sigma Ti_t = \frac{\sum_{j=1}^{j=m} (\mu Ti_t - Tij_t)^2}{m} \tag{10.7}$$

When this work started, a set of bearing problems detection rules were already implemented in the TCMS. The principal reason for the implementation of these rules was train safety. When a certain warning message is recorded in the system during a train trip, maintenance analysis is required, or the train can even come to an immediate stop once the emergency loop opens. The rules are:

- Safety protection rules A. Difference in bearing temperature by axle:

  1. When the temperature difference between the four bearings of the same axle is greater than 25 °C ($Tmax_j$ - $Tmin_j \geq 25°$) and this condition is maintained for more than 1 min, a message A1 is displayed on the PS and stored.

2. When the temperature difference between the four bearings of the same axle falls below 25 °C ($Tmax_j$-$Tmin_j$ <25°) the message A1 is deactivated.
3. When the temperature difference between the four bearings of the same axle is greater than 40 °C ($Tmax_j$-$Tmin_j \geq$ 40°) and this condition is maintained for more than 1 min, and a message A3 is displayed on the PS, it is stored in history, and the emergency loop will open.
4. When the temperature difference between the four bearings on the same axle falls below 40 °C ($Tmax_j$-$Tmin_j$ <40°) the message A3 is deactivated, and the loop is closed.

- Safety protection rules B. Absolute bearing temperature by bearing

5. When the temperature of a bearing reaches a value, $Ti_j \geq 110$ °C and it is maintained for more than 1 min, the loop opens and the train stops. A diagnostic message B5 is displayed on the PS, it is stored in history and the emergency loop will open.
6. When the temperature of a bearing reaches a value, $Ti_j \geq 110$ °C and it is maintained for more than 1 min, the opening of the loop occurs and the train stops. When the temperature of the bearing drops sufficiently, Ti_j <110° deactivates the message B5 and the loop is closed.

The new CBM proposed rules, for predictive maintenance purposes only, without direct implications on train control, were as follows:

- Prediction of abnormal bearing temperature behavior rules C.

7. When the temperature of a bearing Tij_t is outside the interval $\mu Ti_t \pm 2 \cdot \sigma Ti_t$ and it is maintained for more than 1 min, a warning W7 will be logged by the TCMS.
8. When $TS_t \geq TS_{TH1}$ and $AE(Tij_t) \geq AE_{THi}$ and it is maintained for more than 1 min, a warning W8 will be logged by the TCMS. Where AE_{THi} and TS_{TH1} are thresholds considered for the rules.

Figure 10.2 shows the possibilities provided by the C rules. Variables determining these rules are presented during a long train trip covering the period June 20th 16:00–June 21st. 18:00 h. As the reader can notice, no W7 warning message was logged, while W8 message would be logged since (approximately depending on rules thresholds as will be discussed in next Section) June 20th 19:00 almost permanently until 8:00 h of the 21st. No A1, nor A3, nor B5 messages were logged (in fact rules A and B were far to be active). This is a clear example of how it is expected that C8 rule overperforms C7 statistical rule and shows potential improvements in early failure detection in bearings.

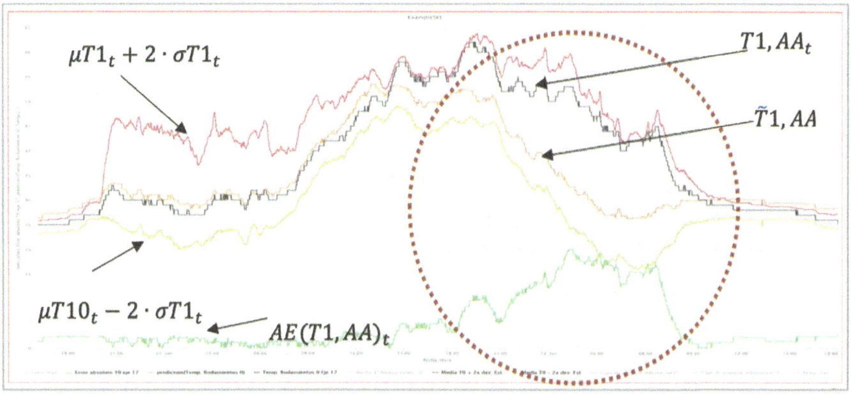

Fig. 10.2 Variable determining C rules in Axle Bearing (1, AA), train TRA, with TS_{TH1} = 50 km/h

The performance of these potential improvements in early failure detection must be tested. To do so, the following process was accomplished:

1. Data sets selection including periods of normal and degraded conditions
2. Descriptors precise plot and thresholds selection
3. Rule expected performance assessment

In order to do the design of the CBM plan rules, a careful review of train bearing data sets was carried out, and a three data sets were selected by the train OEM engineers and bearing technologists. Within those datasets they determined the periods when they would consider the bearing to be in good conditions before degradation, during a degraded state, and in good conditions again after the preventive replacement. In all cases, bearings were replaced ASAP, after the first single message A1 was logged in the system. No A3 nor B5 messages were never logged. Bearings were replaced, not repaired, and the failure mode description was simply "bearing overheating".

10.5 Discussion of Results

The precision of the C8 rule depends on its threshold values. The lower the value of the threshold in absolute error (AE_{THi}) the higher the possibility of false positives (false abnormal temperature behavior), and the higher the possibility of true negatives, and therefore the lesser precision and higher negative predictive value of the rule.

When bearings are in good conditions, Fig. 10.3 shows interesting data coming out of the prediction AE records when using the first data set included in Chap. 6. Practically, $AE(Ti, AA_t) \leq 10\,°\mathrm{C}\ \forall i$, and for any speed regime of the train. $AE(T1, AA_t)$ Distribution is presented in the right-hand site in the same figure to appreciate this fact.

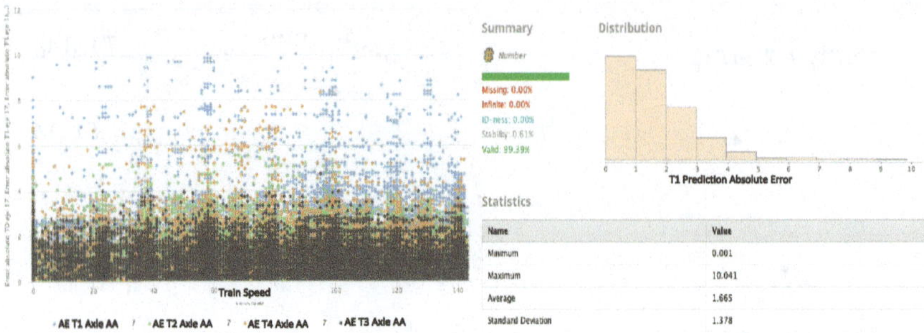

Fig. 10.3 Distribution of EA in T1 bearings in Axle AA (right), and Scatted diagram of all four axle bearing temperature records for different TS. Bearings in good condition

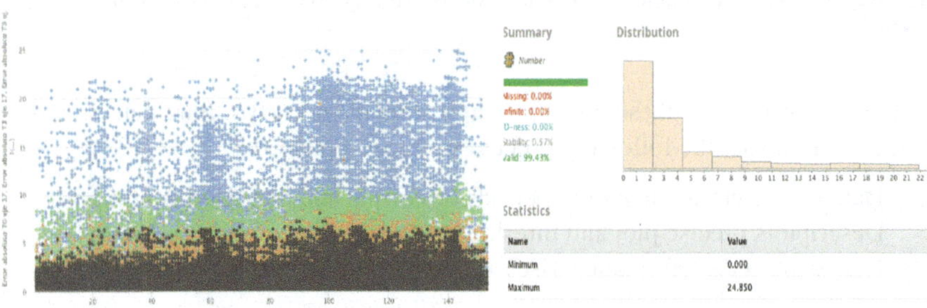

Fig. 10.4 Distribution of EA in T1 bearings in Axle AA (right), and Scatted diagram of all four axle bearing temperature records for different TS. Bearings in bad condition

When bearings are in degraded conditions, Fig. 10.4 shows interesting data, with many bearing T1 points with $10\,°\mathrm{C} \leq AE(T1, AA_t) \leq 25\,°\mathrm{C}\ \forall i$. $AE(T1, AA_t)$ distribution is presented on the right-hand site in the same figure, to appreciate this fact. Moreover, these points are present for any speed regime of the train, but clearly a higher density is found when train speed is higher than 90 km/h. $AE(T1, AA_t)$ distribution is presented in the same figure to appreciate this fact.

To illustrate the difference in AE data distribution when the bearing is in good conditions versus when it comes to a degraded state, Fig. 10.5 represents the temperature prediction AE distribution according to the Rule C 8, for T1, Axle AA in Train TRA period 2 and 3, with $TS_t \geq 90$ km/k.

In the relevant literature, different methods can be found to evaluate the performance of a classification system, one of the most common being the confusion matrix [25], where, Precision is ratio of the number of correctly classified positive examples and the number of predicted positive examples. It is a measure of correctness in positive prediction. Accuracy is the proportion of correct predictions (see Fig. 10.6).

As mentioned above, the number of records considered for error estimation in the data set is fixed by the train speed threshold. Nevertheless, all cases studied show high performance in precision if a rule is selected with an AE_{TH1}=10 ℃.

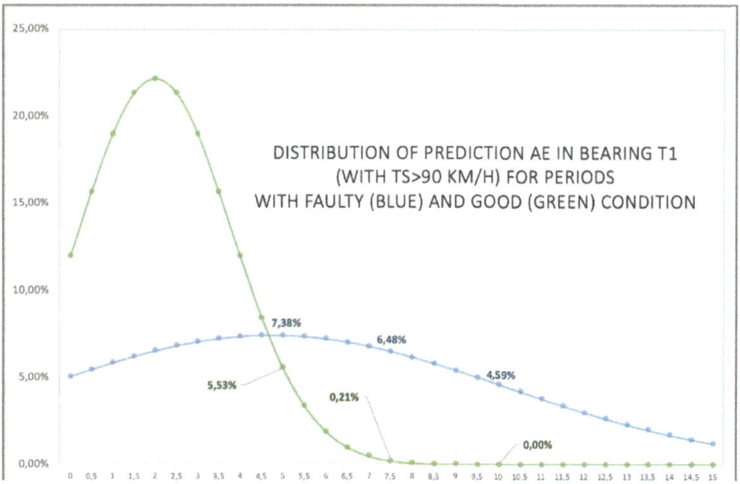

Fig. 10.5 Distribution of EA for good (1, 3) and degraded condition (2) periods, $TS_t \geq 90$ km/h

		Predicted Class: Positive	Predicted Class: Negative	
Actual Class	Positive	True Positive (TP)	False Negative (FN)	Sensitivity $\frac{TP}{(TP+FN)}$
	Negative	False Positive (FP)	True Negative (TN)	Specificity $\frac{TN}{(TN+FP)}$
		Precision $\frac{TP}{(TP+FP)}$	Negative Predictive Value $\frac{TN}{(TN+FN)}$	Accuracy $\frac{TP+TN}{(TP+TN+FP+FN)}$

Fig. 10.6 Confusion matrix and metrics definition

Using values in Fig. 10.5, confusion matrices, as in Fig. 10.6, are built and the prediction rule performance metrics calculated. An example is presented in Table 10.2 for the T1 bearing. The rule performance is examined for two different train speed thresholds 10 and 90 km/h to check differences.

The most relevant conclusion of this Section is the possibility to predict abnormal bearing temperature behavior with maximum precision and specificity, when selecting the Absolute Error descriptor with a threshold of 10 ºC in the rule, and regardless the speed of the train. No relevant results regarding the accuracy of the rule were reached when increasing the threshold for train speed up to 90 km/h.

Table 10.2 Prediction rules metrics for the case of Bearing **T1, 17/T15**

Bearing T1, AA/TRA rule:	Precision (%)	Negative predictive value (%)	Sensitivity (%)	Specificity (%)	Accuracy (%)
$TS_t \geq 10$ km/k, AE_{TH1}>10 °C	100,00	34,13	15,05	100,00	41,01
$TS_t \geq 90$ km/k, AE_{TH1}>10 °C	100,00	37,59	18,35	100,00	45,27

For the case study, once this option could be implemented easily, the final thresholds selected were: AE_{TH1} >10 °C and $TS_t \geq 90$ km/k.

Once precision of the selected rule is high, positives found announced (except for abnormal sensor behavior, situation controlled with other rules in the TCMS) clearly the start of the bearing damage periods. Information obtained about the duration of these periods for the three data set analyzed is presented in Fig. 10.7, where can be appreciated that Period 2 in Train TRA for T1 lasted 49.908 km, for T4 in Train TRB lasted 108.455 km and T2 for Train TRA lasted 75.526 km.

It can be expected that, after the CBM plan is introduced, the amount of damaged bearing data obtained will be properly screened, and further accumulation and analysis of this type of data will help to develop precise predictive models for the remaining useful life (RUL) of the bearing (prognosis) [26]. Although it is still early to have probabilistic estimations about the interval P-F (interval from potential failure to functional failure) of the bearings, initial results are promising in terms of the possibility of enlarging PM intervals of these type of bearings up to 30.000 km, which is a current business demand. Notice that if a damaged state is detected, according to the result obtained from the three data sets, the preventive replacement could be schedule for the next PM interval with very few risks in train operations affection.

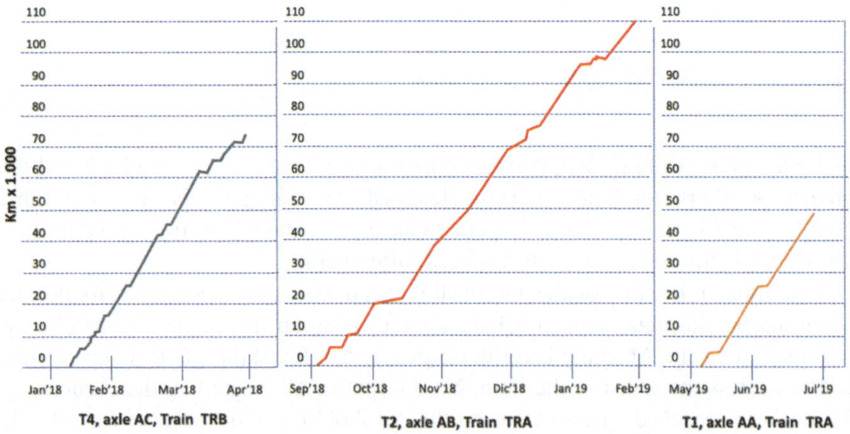

Fig. 10.7 Accumulation of kilometers in period 2 (damaged) for each replaced bearing

References

1. Shin YC, Xu C (2017) Intelligent systems: modelling, optimization, and control. CRC Press, Boca Raton, FL, USA
2. IIC-Industrial Internet Consortium (2016) The industrial internet of things, volume B01: business strategy and innovation framework (IIC: PUB: B01: V1. 0: PB: 20161115). http://www.iiconsortium.org/pdf/Business_Strategy_and_Innovation_Framework_Nov_2016pdf. Accessed 8 Nov 2016
3. Barton D, Court D (2012) Making advanced analytics work for you. Harv Bus Rev 90:78–83
4. Chau KW (2019) Kwok-wing Chau integration of advanced soft computing techniques in hydrological predictions. Atmosphere 10:101. https://doi.org/10.3390/atmos10020101
5. Fayyad UM, Piatetsky-Shapiro G, Smyth P (1996) Knowledge discovery and data mining: towards a unifying framework. KDD 96:82–88
6. Clarke R (2016) Big data, big risks. Inf Syst J 26:77–90
7. Kumar M, KP A (2019) Rolling Bearing Damage. Recognition of damage and bearing inspection. https://doi.org/10.13140/RG.2.2.33971.17447
8. Mitchell TM (2006) The discipline of machine learning. Mach Learn. https://doi.org/10.1080/026404199365326
9. Barke D, Chiu KW (2005) Structural health monitoring in the railway industry: a review. Struct Health Monit. https://doi.org/10.1177/1475921705049764
10. Ferroni F, Klimmek M, Aufderheide H, Laia J, Klingebiel D, Davidich M (2018) Data driven monitoring of rolling stock components. In: Lecture notes in networks and systems, vol 15. Springer, pp 1003–1013. https://doi.org/10.1007/978-3-319-56994-9_68
11. Schaeffler Technologies GmbH & Co. KG (2010) Rolling Bearing Damage Recognition of damage and bearing inspection. (S. Technologies & G. & C. KG, Eds.) (WL 82 102/). Industriestrasse 1– 3 91074 Herzogenaurach Germany: Schaeffler Technologies GmbH & Co. KG. https://www.schaeffler.de/content.schaeffler.de/en/news_media/media_library/index.jsp?tab=mediathek-pub&uid=114073&subfilter=app:dc;language-vid:167;language-pub:167;mediatyp-pub:all;referencetyp-pub:0
12. Nagumo Y, Tanifuji K, Imai J (2010) Basic study of wheel flange climbing using model wheelset. Nihon Kikai Gakkai Ronbunshu, C Hen/Trans Jpn Soc Mech Eng Part C. https://doi.org/10.1299/kikaic.74.242
13. Hamadache M, Jung JH, Park J, Youn BD (2019) A comprehensive review of artificial intelligence-based approaches for rolling element bearing PHM: shallow and deep learning. JMST Adv. https://doi.org/10.1007/s42791-019-0016-y
14. Kou L, Qin Y, Zhao X, Fu Y (2019) Integrating synthetic minority oversampling and gradient boosting decision tree for bogie fault diagnosis in rail vehicles. Proc Inst Mech Eng Part F J Rail Rapid Transit 233(3):312–325. https://doi.org/10.1177/0954409718795089
15. Li H, Parikh D, He Q, Qian B, Li Z, Fang D, Hampapur A (2014) Improving rail network velocity: a machine learning approach to predictive maintenance. Transp Res Part C Emerg Technol. https://doi.org/10.1016/j.trc.2014.04.013
16. Shao H, Jiang H, Li X, Liang T (2018) Rolling bearing fault detection using continuous deep belief network with locally linear embedding. Comput Ind. https://doi.org/10.1016/j.compind.2018.01.005
17. Basheer IA, Hajmeer M (2000) Artificial neural networks: Fundamentals, computing, design, and application. J Microbiol Methods. https://doi.org/10.1016/S0167-7012(00)00201-3
18. Zhang G, Eddy Patuwo B, Hu Y, M. (1998) Forecasting with artificial neural networks: the state of the art. Int J Forecast. https://doi.org/10.1016/S0169-2070(97)00044-7
19. Berry MJA, Linoff G (2004) Data mining techniques: for marketing, sales, and customer relationship management. Wiley
20. Breiman L (2001) Random forests. Mach Learn 45(1):5–32. https://doi.org/10.1023/A:1010933404324

21. Hill T, Lewicki P, Lewicki P (2006) Statistics: methods and applications : a comprehensive reference for science, industry, and data mining. StatSoft. https://books.google.es/books?id=TI2TGjeilMAC
22. Borovicka T, Jirina M, Kordik P, Jiri M (2012) Selecting representative data sets. In: Advances in data mining knowledge discovery and applications. https://doi.org/10.5772/50787
23. Guillén AJ, Crespo A, Gómez JF, Sanz MD (2016) A framework for effective management of condition based maintenance programs in the context of industrial development of E-Maintenance strategies. Comput Ind 82:170–185. https://doi.org/10.1016/j.compind.2016.07.003
24. Gulledge T, Hiroshige S, Iyer R (2010) Condition-based Maintenance and the product improvement process. Comput Ind. https://doi.org/10.1016/j.compind.2010.07.007
25. Koutroumbas K, Theodoridis S (2008) Pattern recognition, 4th edn. Burlington. ISBN 978-1-59749-272-0. Accessed 8 Jan 2018
26. Zhang B, Zhang S, Li W (2019) Bearing performance degradation assessment using long short-term memory recurrent network. Comput Ind. https://doi.org/10.1016/j.compind.2018.12.016

Chapter 11
Asset Condition and Operations Efficiency

11.1 Introduction

The identification and prediction of potential failures can be improved using advanced analytics to search proactively and reduce risk to improve efficiency in energy generation. Algorithms, such as machine learning, are now quite extended in renewable energy control systems. These kinds of facilities are characterized by the presence of a great number of sensors feeding the SCADA systems (supervisory control and data acquisition systems), usually very sophisticated systems including a control interface and a client interface (the plant's owner, distribution electric network administrator, etc.). Power and energy production measures are two of the most important variables managed by the SCADA. As principal system performance outputs, they can be exploited through data mining techniques to control system failures, since most of the systems failures directly affect the output power and the energy production efficiency [1].

11.2 Case Study. Efficiency in Renewable Energy Generation

Predictive analytics can be applied to renewable energy production, as one of the most promising fields of application of these techniques [1]. The stochastic nature of these energy sources, and the lack of a consolidated technical background in most of these technologies, make this sector very susceptible to the application of intelligent optimization modelling techniques. The referred stochastic nature is determined by circumstances in the generation sources, but also by the existing operational conditions. That is, the natural resources have variations according to weather with a certain stationarity but with difficulties in forecasting behaviors. In addition, depending on the operational and environmental stresses in the activities,

A. Crespo Márquez, *Digital Maintenance Management*, Springer Series in Reliability Engineering, https://doi.org/10.1007/978-3-030-97660-6_11

Table 11.1 References of DM techniques analyzed

Techniques	References
Data mining	[3, 5–7]
Artificial neural networks	[1, 8–14]
Support vector machine	[1, 34–37]
Decision trees	[14, 21, 26, 38–42]

they will be more likely to fail. Consequently, the analysis of renewable energy production must consider adaptability to dynamic changes that can yield results [2].

In the future, assuming more mature application scenarios, the selection of predictive analytics techniques will likely be crucial to generating well-informed decisions.

Accepting this challenge, a review of the literature, the selection of techniques and a benchmark of their results are presented in this Chapter. According to the previous literature, most representative techniques [3, 4] are presented and applied to a case study in a photovoltaic plant (see other examples where these techniques were applied in Table 11.1).

In general terms, the results obtained using machine learning to follow and predict PV critical variables, for example solar radiation [5], are good enough to use as inputs in decision-making processes, such as maintenance decisions [1]. However, as will be appreciated in this Chapter, not all of the techniques have the same maturity level as techniques to predict the yield of a PV plant, some of them should be studied in greater depth in the coming years [6].

11.2.1 Reviewing the Use of Predictive Techniques

A classification of predictive techniques in the photovoltaic sector is presented in [4]. These results show how data mining techniques are becoming increasingly relevant, since they represent 61% of the total of the studies. Another interesting classification study is included in [7]. Making a general and in-depth presentation of different predictive techniques is a very interesting task that goes well beyond the aims of this Chapter. Figure 11.1 presents a basic classification of techniques that are going to be compared in this Chapter by applying them to the same case study. In the following paragraphs, a brief literature review introducing these techniques is included.

Table 11.1 summarizes references employed in the paper corresponding to those techniques analyzed.

- Artificial neural networks (ANN)

  In estimations concerning renewable energies, ANN techniques are widely utilized and, more particularly, the field of photovoltaic systems has been continuously developing them in recent years [8–10]. There are various ANN

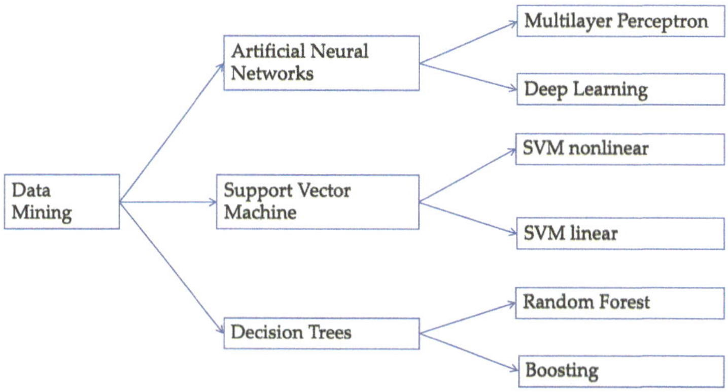

Fig. 11.1 Classification of DM techniques included in this paper

models, and a particular architecture widely extended is multilayer perceptron (MLP) [15].

In [11] a study is presented to obtain with greater precision the production of electrical and thermal energy from a photovoltaic and thermal concentration system, using a neural network (multilayer perceptron) to predict solar radiation and irradiance. In a maintenance application, in [1] the authors go further in their study using the predictive model obtained with the multilayer perceptron neuronal network trained with the backpropagation algorithm to anticipate the occurrence of failures and, thus, improve the efficiency of the final production.

Deep learning neural networks are multilayer and feedforward neural networks that consist of multiple layers of interconnected neuron units with the aim of construing better level features, from lower layers to a proper output space. The application of deep learning techniques provides a fairly accurate prediction in renewable energies, and the authors [13] use a deep learning model to try to mitigate the risks of uncertainty in the production of a wind farm, testing this model in several wind farms in China. The result obtained with this technique improves those obtained with others and avoids the uncertainty of energy production due to climate change. As for hydrological predictions, there are few studies using deep learning techniques, and the authors present their results [14], and while they are just a beginning, the results are promising.

- Support vector machine (SVM)

Inside the supervised machine learning techniques, support vector machines (SVM) [16] are properly related to classification and regression problems, representing in a space two classes, maximally separated through a hyperplane with high dimensionality (defined as a vector between two points of each class), that permit the classification of new data in one or both classes. Regarding the application of SVM techniques, the authors [17] present a study on the prediction for cooling of an office building in Guangzhou, China. For this purpose, they use the comparison of different neural network techniques (NNBR, NRBR, NRBR, NRBR) and NSRV, based on the results obtained in each of them from the mean square error and the relative mean (RMSE and MRE). This model of artificial intelligence (SVM) is, in this case, the one that provides the best result, obtaining a high precision in the hourly prediction of the building's cooling and significantly improving the results of the neural networks.

Likewise, there are numerous references for the application of this technique in the renewable energy sector due to the good results obtained with them. The authors [18] use this technique to predict the average daily solar radiation using air temperature and analyzing the result obtained by the highest correlation coefficient (0.969) and the lowest mean square error (0.833), which shows the promise of this new technique compared to traditional methods. The authors [18] attempt to predict the production of a wind farm in the short term, through wind speed, wind direction and humidity. They compare SVR techniques (multi-scale support vector regression) with a multilayer perceptron neural network, obtaining better results with SVR due to its speed and robustness. Regarding hydrological forecasting, there are also references, such as the [19], that use the RSVMG (recurrent support vector model) technique to predict the volume of rainfall during the typhoon season in Taiwan. Shi in [20], for their part, use this technique to predict the output of a photovoltaic installation in China and verify the result through the RSME. Although it is a relatively recent technique, the results obtained are very promising and encourage further research in this field.

- Decision trees (DT)

RF (Random Forest) is one of the most recent techniques that will be applied in our case study and has obtained very good results. Some examples are presented below:

- Elyan, E. in [19] uses the RF technique to classify data, demonstrating that it is a very accurate method of classifying and obtaining results that improve accuracy over other techniques.
- Lin, Y. in [21] uses RF to improve the prediction of wind production in the short term, which is complicated by the stochastic nature of the wind and using the effects of seasonality. RF modelling obtains accurate results in this case.
- Moutis, P. [22] presents two applications of decision tree techniques: the planning of organized energy storage in microgrids and energy control within a PC through the optimal use of local energy resources, demonstrating through a case study the feasibility of this technique.
- Ren, L. [23] use the DT technique to predict surface currents in a marine renewable energy environment in Galway Bay. The results obtained are very promising, obtaining a correlation coefficient higher than 0.89.

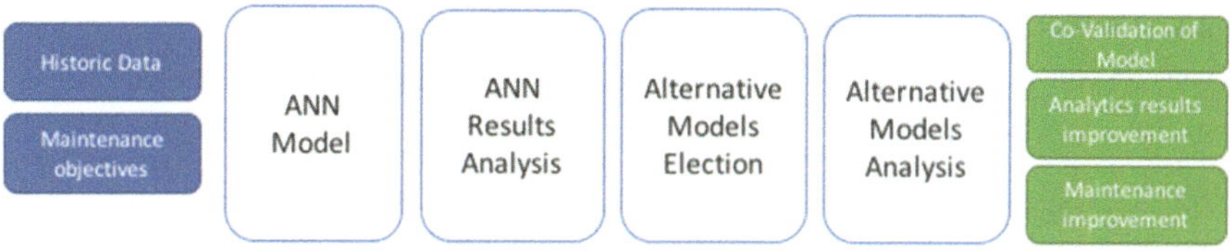

Fig. 11.2 Methodology for using alternative DM techniques

The following figure (Fig. 11.2) shows the methodology that will now be applied for the selection of techniques whose behavior pattern best suits the productive model of a given facility.

11.2.2 Case Study Development

The methodology will be applied to a photovoltaic installation with 6.1 Mw of rated power that is in Córdoba (Spain) and has been in operation since 2008. This facility is divided into 61,100 kW solar orchards. Applying the study on three of these orchards it has been verified that the results in all three are analogous, so we set out only one of them. Tables 11.2, 11.3 and 11.4 show the information taken for the study:

Results obtained for each technique for different techniques are shown below (Table 11.5), as well as the different transformations made (different ways to normalize variables and to estimate parameters).

This model shows the importance of variables in the result, all of them are valid and necessary. The higher the percentage, the higher the importance variable (see Table 11.6). The prediction error based on %INC_MSE is estimated by out-of-bag

Table 11.2 Temporary period for data collection for study

Start date	End date	Data collection	From	Until	Frequency
01/06/2011	30/09/2015	Hourly	8:00	17:00	10 daily data for each variable

Table 11.3 Selected variables/data for training and validation

Inputs variables	Outputs variables	Selected values	Training set percentage	Testing set percentage
Outdoor temperature, radiation, inside inverter temperature, operation hours	Time, production	In the absence of failures	75%	25%
			Same criteria for all techniques in order to establish the same comparison environment	

Table 11.4 Collected data in the study (treated and validated)

Outdoor temp	Radiation	Indoor temp	Operating time	Production
303.25	490.3	312.1	9900	52
313.25	756.0	311	9901	74
319.25	860.8	314.7	9902	80
323.25	901.8	313.9	9903	82
325.25	918.0	315.5	9904	83
327.98	990.3	316.8	25,716	81
320.77	520.0	315,6	25,717	53
311.43	454.5	317,1	25,718	39
305.98	777.3	317,8	25,719	66

Table 11.5 Sample Results of the different techniques

Models analysis	Data	Correlation coefficient	RMSE
Multilayer perceptron	Test	0.886	9.64
	Training	0.897	9.15
Deep learning	Test	0.839	11.5
	Training	0.855	10.93
SVM nonlinear	Test	0.881	10.01
	Training	0.894	9.46
SVM linear (Lib linear)	Test	0.836	11.57
	Training	0.848	11.1
Random forest	Test	0.909	8.63
	Training	0.916	8.3
Boosting	Test	0.856	10.87
	Training	0.868	10.41

Table 11.6 Importance of the variables

Variable	%INc_mse
tEMP_EXT	49.10
RADIATION	155.63
TEMP_INT	41.60
Operating hours	34.37

(OOB) for each tree and after permuting each predictor variable, until the difference between them has a standard deviation equal to 0.

A point cloud chart (Fig. 11.3) of the predicted (test) production is shown for the model that give us the best solution (Random Forest).

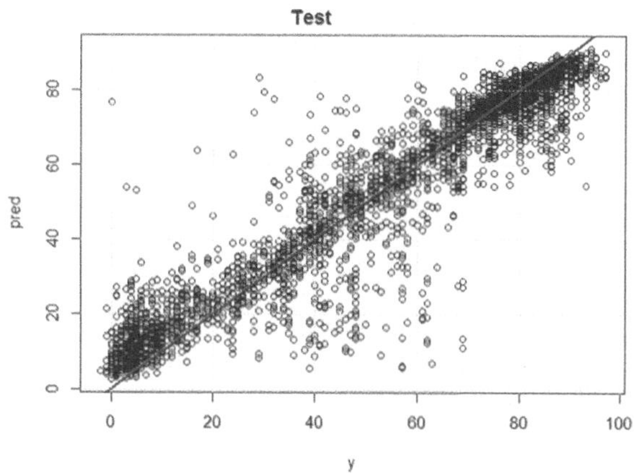

Fig. 11.3 Point cloud (test) Random Forest

11.2.3 Case Study Conclusions

In this Section a methodology to introduce the use of different techniques for energy forecasting and to support condition-based maintenance was followed. These techniques compete for the best possible replica of the production behavior patterns of a renewable energy (PV installation).

In this study a very large sample of data, five years, has been considered and all the models for the different techniques offered very encouraging results, with correlation coefficients greater than 0.82. Coinciding with other referenced authors' results, Random Forest was the technique which provided the best fit, with a linear correlation coefficient of 0.9092 (followed by ANN and SVM). In turn, this technique (RF) gave us as a differential value of the importance of the input variables used in the model, which somehow validates the use of all these variables. In the case study, and by far, the variable resulting with the most affection to production was radiation, followed by the outside temperature, the inverter internal temperature and, finally, the operating hours (which somehow reflects the asset degradation over time). It is important to mention that these results were obtained using different methods to normalize the variables and to estimate parameters.

Future work could be devoted to the validation of these results by replicating the study at other renewable energy facilities to determine how the improvement in ECM and R2 values affects early detection of failures by quantifying their economic value. The implementation of these techniques is feasible today thanks to existing computational capacity, so the effort to use any of them is very similar.

11.3 Case Study. LNG Storage Tank Pump Operation Analysis

As instrumentation and information systems become cheaper and more reliable, CBM becomes an important tool for running a plant or a factory [24, 25]. In automated manufacturing or process plants, condition-based maintenance (CBM) is preferred wherever it is technically feasible and financially viable [26]. Condition monitoring techniques, along with developments in Prognosis and Health Management (PHM) has provided new powerful capabilities for maintenance [27, 28]. These capabilities allow us to treat, efficiently, the new maintenance challenges in modern systems and applications [29, 30].

However, this Section focuses on situations when asset's performance monitoring becomes a very complex problem, for instance when:

- The assets can perform in very diverse operating conditions (sometimes also diverse environmental conditions) over time.
- Assets' condition cannot feasibly be monitored, or simply doing it becomes a complex technical problem with a very troublesome and economically non-viable solution. This difficulty is frequently related to the asset specific functional location.
- Altogether could cause a lack of assets performance control and subsequent loss of expected performance efficiency.

To overcome these problems, practical tools that can be adapted to any potential asset location and operating condition are needed. Tools that have the possibility to control asset performance and reliability, ensuring life cycle expectations according to existing business plans.

This has become an important concern for many industries and utilities with very intensive capitalization in long-lasting assets [31]. These organizations demand innovative, practical solutions, that can somehow replicate the benefit provided by emerging IoT and CPS, but in currently existing assets, contributing to their current assets' Total Cost of Ownership (TCO) and/or Total Value of Ownership (TVO) [32], the latter in case we can somehow model, and quantify, the value provided by the asset along its life cycle.

11.3.1 Using Data Mining Together with Predictive Analytics

In this Section a set of data mining (DM) techniques will be used together with ANN algorithms, to solve the problem of identifying when asset behavior abnormalities appear (ANN), for which operating conditions are appreciated, and to what extent is energy efficiency impacted (DM AR). Then, a proposed process will be presented to deploy the implementation of these techniques.

Artificial Neural Networks (ANN) is the tool selected because of their ability to process information features in non-linear, high-parallelism, fault and noise environments with learning and generalization capabilities [33, 34]. In comparison to other methods, ANNs are well suited to solving problems where explicit knowledge is difficult to specify or define, but where there is enough data [35, 36]. ANN are a very popular tool for prediction and classification problems, but more research is needed to make their implementation more practical in real life applications, taking advantage of existing maintenance engineering tools and management systems [1].

Concerning data analysis tools, several DM techniques exist and, according to the purpose of each study and to the characteristics of the variables under investigation, the accurate selection of the most appropriate one is the first step for the success of the analysis [37]. We propose a classification of the DM techniques, differentiating among models requiring or not a reference value for the analyzed variable. The former group is furtherly detailed, discriminating between descriptive models, namely the ones oriented to defining relationships among objects (e.g., conceptual clustering), and associative models, addressing the relationships among variables (e.g., association rules, model-based reasoning, reason weighting). According to this, among the DM techniques, we selected Association Rules (ARs) mining for two main reasons: firstly, because of its purpose, that is discovering relevant and non-trivial attribute-value relations not immediately recognizable in a massive amount of data [38]. The second one is related to ARs' mining intuitiveness that makes results interpretation instantaneous even for non-domain experts. Also, ARs mining is very popular in several different fields of application.

ARs were first applied to determine the relationships among items frequently bought together [39]. With the growing popularity of such a methodology, the applications to other fields increased as well, further highlighting the versatility of the methodology. For example, in the manufacturing field, ARs mining is applied to relate the root cause to groups of machines in a production process [40] or in predicting process failures [41, 42].

Other ARs mining applications regard the analysis of events co-occurrences or correlations. For instance, they can be used in defining the occurrence of couples of events, as presented in case of trains diagnostics [43]: rules relating sequential events are taken into consideration, to monitor train fleets more efficiently. In a similar manner, process reliability and safety are addressed by several authors: [44] develop a procedure for failure detection on machines belonging to different production lines, by integrating maintenance data; [45], instead, apply the ARs to deploy a Total Productive Maintenance strategy to increase the reliability of the production process. Maintenance policies based on ARs mining for the prediction of components' breakages are also applied in existing contributions related to refinery plants, both involving an active participation of plants operating personnel in parameters setting [46] and automatizing the latter aspect through the resolution of an optimization model considering production constraints [47]. Conversely, some authors showed that the application of ARs mining in power plants has a positive impact for diagnostic purposes but are not a sufficient in predicting failure occurrences [48].

The ANNs and ARs mining approaches are promising methodologies for prediction and classification, but more research is needed in terms of combination of the two. A work integrating the ARs mining and ANNs approaches operational reliability issues is already existing in literature: specifically, the ARs are extracted to define the factors influencing power distribution reliability; then, such factors are given in input to an ANN for the normal behavior modelling of the distribution net [49]. The approach proposed in the current paper is the opposite: indeed, the normal behavior is firstly estimated through an ANN and, when a substantial deviation is noticed, the ARs mining is applied to diagnose whether such a deviation impacts on the efficiency and performance of the asset. In fact, if the ability of ANN models to predict the normal behavior of a system starting from the analysis of the process' variables is indisputable, it might also be useful to understand what the impact that influences the abnormal behaviors of the process is. Such an analysis can be easily carried out through the ARs mining. For this reason, the approach proposed in the current paper regards the combination of the ANN and ARs mining.

11.3.2 An Introduction to the DM AR Technique

In this Section, the Association Rules (ARs) technique is formally defined, and a description of the rule mining algorithm is provided in the following.

Given a set of Boolean data I = {i1, i2, ... in} called items and a set of transactions T = {t1, t2, ... tm}, a transaction ti is defined as a subset of items belonging to I (i.e., an item-set). An Association Rule (AR) is an implication in the form $\Gamma \rightarrow \Delta$, where Γ and Δ are item-sets belonging to I and whose intersection is null (i.e., $\Gamma \cap \Delta = \emptyset$). Γ and Δ referred to as body and head of the rule, respectively. The quality of an AR can be assessed through different metrics; in this study, we apply the most popular ones that are (1) the Support and (2) the Confidence.

(1) Support ($\Gamma \rightarrow \Delta$) = (| Γ, Δ|)/m: the support is the ratio between the number of transactions both containing the item-sets Γ and Δ (|Γ, Δ |) and the total number of transactions (*m*). Hence, it represents the probability of finding a transaction containing Γ and Δ.

(2) Confidence ($\Gamma \rightarrow \Delta$) = (Support ($\Gamma \rightarrow \Delta$))/(Support ($\Gamma$)): the confidence can be considered as a measure of the strength of the rule. Indeed, it is calculated as the support of $\Gamma \rightarrow \Delta$ over the support of item-set Γ. In practice, confidence represents the probability of occurrence of Δ, given the occurrence of Γ, i.e., the conditional probability of Δ, given Γ.

The procedure for mining the ARs from a dataset, starts from the generation of the Frequent Item-sets: the users define a minimum support according to the dataset's characteristics; then, the item-sets appearing in T with a frequency higher than the minimum support are generated through the FP-growth algorithm [50]. Starting from each frequent item-set F, all the rules in form X → Y, such that X U Y = F, are generated.

Since assets' performance monitoring involves the analysis of continuous variables, the discretization of such attributes should be considered. This issue is already addressed in existing literature: indeed, some authors provided guidelines for discretizing such values. For example, [51] propose an entropy minimization-based heuristic to define multiple intervals, while [52] propose the application of an evolutionary algorithm to define ranges automatically. Data could also be clustered before extracting the ARs, even if some relationships can be disregarded (e.g., the support of such rules does not reach the minimum support required). In order not to lose any relevant information, in the energy consumption monitoring field, data is often partitioned according to a specific strategy, e.g., in fixed ranges [53]. Hence, in our application, the discretization of the continuous variables is defined in accordance with the O&M department, so that all the different operating conditions can be mapped. The software RapidMiner (www.RapidMiner.com), an opensource data mining platform, is selected to carry out the analysis. Among the various tools available on the market, RapidMiner is considered to be one of the most user-friendly, because of the graphical interface and the ease of modifying the parameters, since no programming skills are required [54].

11.3.3 The Emerging APM and Reliability Asessment Process

The "Asset Performance Monitoring and Reliability Assessment Process" coordinates and integrates data, engineering tools and human resources (with knowledge, skills, and experience) to achieve the best possible recommendation for assets' operations and maintenance actions. This process also analyzes the different renovation and reinvestments options which are available for the physical asset. All these activities are accomplished considering the different operational and data related constraints and controls, as presented using the Functional Modelling method IDEFØ diagram [55] of the process in Fig. 11.4.

This process is a continuous improvement process responsible for leading and controlling the knowledge captured concerning the behavior of the asset (observed properly when integrating O&M Data [I1], with process [I2] and condition monitoring data [I3]), adapting permanently operations and maintenance actions. The process provides consistency to major maintenance actions and assets replacement options ensuring proper AM orientation according to predictable and/or unpredictable circumstances that may take place.

The general process in Fig. 11.4 can be split into four different sub-processes, named modules in this paper, as illustrated in Fig. 11.5.

The first module is Data Processing, responsible for the preparation of valid data, quality tested (O1), for the subsequent processes, as well as for the normalization of this data (O2). The second module is the one to provide an algorithm that can identify deviations of the asset from its expected behavior under normal operating conditions. These deviations are presented in the form of prediction error (O3). The third module deals with the measurement of the deviation, obtaining the relationship

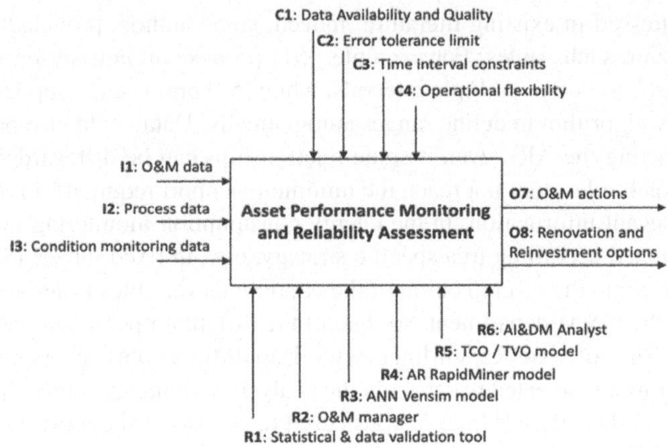

Fig. 11.4 Process general representation

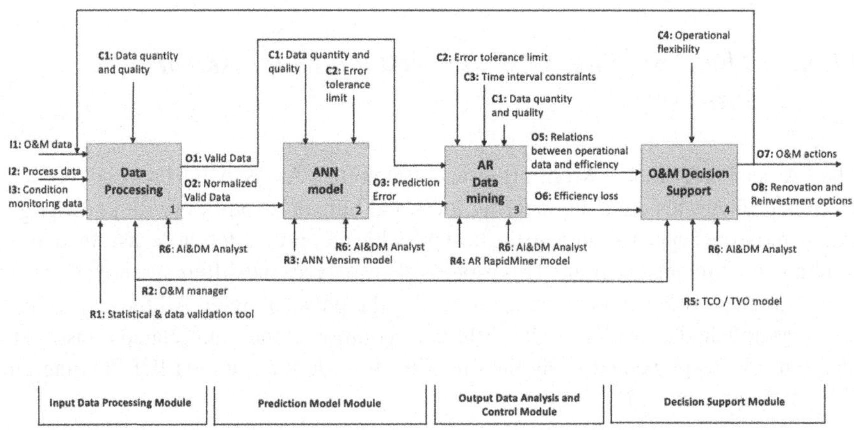

Fig. 11.5 Level-1 representation of the asset performance monitoring and reliability assessment process

between operational data and efficiency (O5) and then calculating the efficiency loss observed for a period (O6). Finally, the last module takes this information to adjust O&M actions properly (O7) and to re-evaluate asset major interventions (O8) accordingly.

The reader is addressed to [56] to see each one of these processes explained in detail, going down to level 2 in the IDEFØ diagram presented (in that paper Functional Modeling method IDEFØ with three hierarchical series of diagrams is used). In addition, the application of each phase of the process is illustrated with a practical example.

For this Section the example of a process pump located inside a cryogenic LNG storage tank has been selected. The pump has the following features:

- A unique design allowing the pump to be installed inside the tank in a vertical column to remove the possibility of major tank leakage due to a pipe or connection problem.
- The pump and motor units are submerged, and the column acts as a guide to seat the pump during installation and performs as the discharge pipe from the pump.
- The plant information system shows operating conditions of the pump to detect any potential problem.
- The pump has very different possible operating conditions (resulting from very different tank levels, flow, pressure, LNG density, operating hours of the pump, etc.).
- It is very troublesome to establish a clear judgement about potential pump malfunctioning. Additionally, condition monitoring capabilities are limited to previously mention operating variables (for instance, no items vibration, bearing temperatures, motor temperatures are received).

In this example, the scope of the analysis was fixed as follows:

- To generate a process and tool for the prediction of anomalies in the operation of the pumps with complex operation and supervision regime.
- To establish the guidelines for the practical implementation of the models and methodology
- To generate enough information to prepare a Business Case for the company to implement the process if there is a demonstrated payback to the business.

Some other secondary or complementary objectives were:

- To identify the operating regimes of the equipment so that anomalies are recorded.
- To obtain the loss of performance of the equipment when producing the anomalies detected.

11.3.4 Presenting the Process Results

The following table shows all available variables and their origin (Table 11.7):

Notice that to improve the choice of variables, it is advisable to use statistical techniques to measure the degree of dependence (see R1 in previous Figures). For the example presented in this paper, the variables that have been selected and that will follow the next phases of the process are those shown in Table 11.8.

This Table is the result of applying statistical techniques; in particular, it has been studied using a covariance matrix and correlation coefficients, the dependence between variables [57]. The analysis of the results assumes that flow, output pressure, output temperature and Hrs. since the last overhaul are indispensable for the model, and that other variables such as fluid density, tank level and intake temperature are secondary, and they could be left out of the model if sufficient data is not available for power consumption prediction modelling.

Table 11.7 Asset case study definition, data source, for the process pump example

Variable	Corrective and preventive maintenance	Fluid density	Flow	Intake temperature & output temperature	Output pressure	Tank level	Power consumption	Operation modes	Hours since last Overhaul
GMAO	√								
LNG quality registration		√							
Plant info. system			√	√	√	√	√		
Shift superintendent book								√	
Pump hour meters									√

Table 11.8 Sample covariance matrix and correlation coefficients, to identify the dependence between variables

Covariance	Flow	Output pressure	Power consumption	Intake temperature	Output temperature	Tank level	Fluid density	Hours since last overhaul
Flow	0.04100							
Output pressure	– 0.01438	0.01096						
Power consumption	0.03624	– 0.01337	0.03503					
Intake temperature	– 0.00531	– 0.00167	– 0.00562	0.014173				
Output temperature	– 0.01038	0.00208	– 0.01078	0.00905	0.02344			
Tank level	0.00565	0.00538	0.00543	– 0.01025	– 0.00348	0.02418		
Fluid density	– 0.00457	0.00241	– 0.00281	– 0.00077	– 0.00340	– 0.00315	0.01663	
Hrs since last overhaul	– 0.00844	0.00100	– 0.00922	0.00279	0.00428	– 0.00392	0.01136	0.08351

Table 11.9 Period selection

Hours	Period before OH 1		Period before OH 2		Period before OH 3	
Operating hours	From 2.375	To 16.147	From 0	To 5.173	From 0	To 94
Valid data points	10.082		4.857		49	

In Table 11.9, the periods selected for the case study are shown. It is divided into three periods; the passage from one period to another is marked by an overhaul (OH) maintenance activity. The data recording frequency has been every calendar hour.

For those periods, as part of data preparation for machine learning, the data for variables was normalized to change the values of different variables in the dataset to a common scale, without distorting differences in the ranges of values. For machine learning, every dataset does not require normalization. It is required only when features have different ranges. Each variable is normalized between 0 and 1, making use of the technical information collected in the first step of the procedure.

For example, if a variable is by design limited between 250 and 700 m3/h, the values that are outside the range are not considered for the study, but the values within the range will be normalized between 0 for values close to 250 m3/h and 1 for values close to 700 m3/h. The maximum (Vmax) and minimum (Vmin) values for each of the variables selected for the case study are shown in Table 11.10.

The prediction model selected was a continuous time ANN simulation model that has demonstrated to be easy to adapt for prediction accuracy and self-learning capabilities.

Regression results for the ANN are presented in Table 11.11.

Table 11.10 Values for data normalization

Variable	Vmin	Vmax
Flow (m^3/h)	250	700
Intake Pressure (kgf/cm^2)	6	14
Intake Temperature (TAsp) (°C)	−160.5	−150
Impulsion Temperature (TImp) (°C)	−160.5	−153.5
Tank Level (Nivel T) (mm)	0	50,000
LNG Density (Dens GNL) (kg/m^3)	430	465
Hrs since last Overhaul (h)	0	16.128
Power (Kwh)	200	350

Table 11.11 Artificial Neural Network (ANN) algorithm results for the regression problem (with 20 N/HL)

Metric	Training	Before the overhaul 1	After the overhaul 1
RMSE	0.00030	0.00043	0.00222
MAE	0.01567	0.02297	0.05447
MAPE	0.02807	0.03963	0.12195

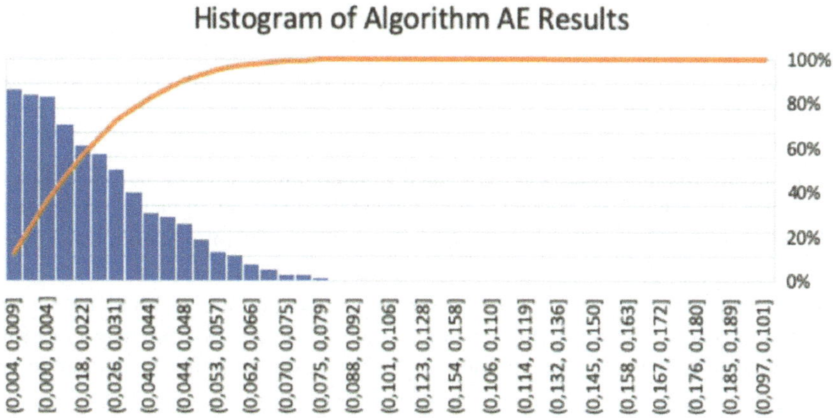

Fig. 11.6 ANN results precision (X axis) after training, for the period before the OH 1

The precision of the algorithm results can be observed in Fig. 11.6, where a histogram of absolute values for the results of prediction before the overhaul of the pump is presented.

Before concluding this phase and although our results are promising, we realize that the selection of a given technique, like ANN, should not be done without discussing whether we are delivering results that can be benchmarked to those provided by other available advanced machine learning techniques for predictive analytics. This certainly supports the tool selection and provides more elements for a well-informed decision for the management of assets [36].

This was a good exercise to test the prediction quality of the ANN algorithm, this fact as well as the maturity in the implementation of this technique within the reliability and assets management organization and some other practical implications, made the selection of the ANN process reasonable. See Table 11.12 for these results.

Input data processing unit and the prediction error are input in the AR data mining activity. The aim of this step is to study the relations among the operating conditions

Table 11.12 Benchmarking the result of the different predictive analytics techniques

Model	Data set	RMSE
ANN model	Before the overhaul 1	0.00043
	After de overhaul 1	0.00222
SVM nonlinear	Before the overhaul 1	0.00045
	After de overhaul 1	0.00230
SVM linear (Liblinear)	Before the overhaul 1	0.00052
	After de overhaul 1	0.00269
Random forest	Before the overhaul 1	0.00038
	After de overhaul 1	0.00201

and the efficiency of the system under investigation (O5) and checking any loss of efficiency (O6).

In our application, the variables selected are pressure, flowrate, impulsion temperature, aspiration temperature, tank level, and density; all of them are related to the efficiency of the system.

The discretization process is performed in accordance with the O&M manager (R2). Indeed, a trade-off is needed in this step: on the one hand, in fact, the discretization ranges must be small enough to represent a specific operating condition. On the other hand, instead, we wish to have the greatest interval size, so that the system frequently operates inside it. In other words, if the intervals are too small, the system could work inside each of them very rarely and, consequently, we may not profit from the data. In the next table (Table 11.13) we report the minimum and maximum values of the valid data and the size chosen for each interval.

Concerning AR mining results: the discretized variables range represent the input of AR mining stage; quality of the results depends on data quality and availability (C1) and the temporal constraints considered (C3). The resources necessary to deploy the data mining procedure are O&M manager (R2) and the AI&DM analyst (R6), together with AR RapidMiner model (R4). According to the methodology explained in Sect. 3, the rules having values of the efficiency as head (Δ), and values of all the other selected variables as body (Γ) must be analyzed. Each AR is associated with a support and a confidence: the former represents the percentage of time units over the total in which the operating conditions and the efficiency have assumed the values expressed by the rule. The latter, instead, indicates the probability of having the efficiency expressed by the rule, given the values of the operating variables. For example, in Table 11.14, some of the rules extracted (O5) are reported. Column 1 to 6 (Flow rate, LNG Density, Tank level, Pressure, Intake Temperature, Impulsion Temperature) contain the body of the rule, while the seventh is the head (Efficiency); column 8 and 9, instead, respectively report the support and the confidence associated with each rule. The rules exemplified in Table 11.9, should be interpreted as follows (let us consider, for instance, the first row): when the operating conditions of the system are characterized by a flow rate included in the range 478–508 m^3/h, a density between 450 and 455 kg/m^3, the tank level between 5,900 and 6,900 mm, the pressure inside the tank is between 10.5 and 11.5 kgf/cm^3, the aspiration temperature (TAsp)

Table 11.13 List of variables, minimum values, maximum values and interval sizes

Variable	Min value	Max value	Size
Flow rate [m^3/h]	328	668	30
LNG Density [kg/m^3]	430	465	5
Tank level [mm]	1,900	35,900	1,000
Intake pressure [kg/cm^2]	6.5	13.5	1
Intake temperature [°C]	− 160	− 157	1
Impulsion temperature [°C]	− 160	− 153	1
Efficiency [%]	42	68	1

Table 11.14 Example of the ARs extracted

Flow rate	Density	Tank level	Pressure	TAsp	TImp	Efficiency	Support	Confidence
[478–508]	[450–455]	[5,900–6,900]	[10.5–11.5]	[−159–158]	[−158–157]	[59–60]	0.0006	0.75
[478–508]	[450–455]	[5,900–6,900]	[10.5–11.5]	[−159–158]	[−158–157]	[60–61]	0.0002	0.25

is included in −159 and −158 °C and the impulsion temperature (TImp) is in −158 and −159 °C, in 75% of the cases (confidence = 0.75), the efficiency of the system ranges between 59 and 60%; in the remaining 25% of the cases (see the second row of the Table), instead, the efficiency range is 60–61%. The support associated with the former relation is 0.0006: this value indicates the probability of having the operating conditions values listed before and an efficiency level in 59–60%. Similarly, the rule associating the operating conditions values and the efficiency between 60–61% have a joint probability of occurrence, namely a support, of 0.0002.

Regarding rules comparison: The relations among the operating conditions and the efficiency (O5) mined in the previous step should be compared in order to verify the actual existence of an efficiency loss (O6) and the corresponding probabilities. To this end, we need to select rules presenting analogue values of the operating conditions but belonging to different time intervals, i.e., those identified in the first step of this stage. This activity has to be carried out both by the O&M manager (R2) and the DM&AI analyst (R6). In our study, three different time intervals have been selected and compared, as shown in Table 11.9. The highest number of comparable conditions characterizes TI 1 and TI 2: specifically, 58 common operating conditions have been identified in which the system under investigation works both before and after the overhaul (i.e., the milestone dividing TI1 from TI2). As shown in the following Table (Table 11.15), in almost the totality of the rules there is a loss of efficiency. Comparing the remaining time intervals there are less coincident operating conditions. In general, the efficiency of the TI3 is higher than the one of TI 2 and greater-equal than the one of TI 1.

A more detailed example is reported in Table 11.16. The first row contains the operating conditions and the corresponding efficiency in TI 1, while in the following three rows the analogue conditions and the efficiency values are reported for TI 2. In all cases, a decrease of the efficiency is noticed: indeed, in TI 1 its value is among 64 and 66%, while in TI2, efficiency values range among 60 and 62%. In TI 3, the efficiency increases, returning to the same range as TI 1. The interpretation of these results should be related to the preventive overhaul executed at the end of TI 1. Possibly, an error during the maintenance intervention or the re-installation occurred,

Table 11.15 Efficiency comparison among the identified time intervals

Time Intervals	Efficiency comparison	Coincident operating conditions
TI 1 versus TI 2	Efficiency TI 1 > Efficiency TI 2	54
	Efficiency TI 1 ≤ Efficiency TI 2	4
TI 1 versus TI 3	Efficiency TI 1 ≤ Efficiency TI 3	16
TI 2 versus TI 3	Efficiency TI 2 < Efficiency TI 3	3

Table 11.16 Rules comparison among the identified time intervals

TI	Flow rate	Density	Tank level	Pressure	T Asp	T Imp	Efficiency	Support	Confidence
1	[508–538]	[445–450]	[27900–28900]	[11.5–12.5]	[−160 - − 159]	[−159 - −158]	[64–65]	0.0003	0.5
	[508–538]	[445–450]	[27900–28900]	[11.5–12.5]	[−160 - − 159]	[−159 - −158]	[65–66]	0.0003	0.5
2	[508–538]	[445–450]	[27900–28900]	[11.5–12.5]	[−160 - − 159]	[−159 - −158]	[60–61]	0.0002	0.5
	[508–538]	[445–450]	[27900–28900]	[11.5–12.5]	[−160 - − 159]	[−159 - − 158]	[61–62]	0.0002	0.5
3	[508–538]	[445–450]	[27900–28900]	[11.5–12.5]	[−160 - − 159]	[−159 - −158]	[64–65]	0.0204	0.5
	[508–538]	[445–450]	[27900–28900]	[11.5–12.5]	[−160 - − 159]	[−159 - −158]	[65–66]	0.0204	0.5

compromising the performance of the entire asset and leading to a catastrophic failure at the end of TI 2. Indeed, in TI 3, the normal level of efficiency was re-established.

As presented in this paragraph, the output of the output data analysis and control module are the relations among the operational, process, condition monitoring variables and efficiency and, after the comparison, a measure of the efficiency loss.

It is important to notice that modifying the head of the rule, i.e., replacing the efficiency with another attribute, allows moving the focus of the investigation. For example, it could be interesting to compare the operational variables with the failure modes, to associate different operating conditions with the most likely failure mode and prevent it through specific actions.

At this point of the process, the O&M manager can make decisions concerning possible changes in the normal operation and maintenance of the asset but can also select a better strategy for the overhaul and reinvestment in the asset or its future replacement. These possibilities are constrained by the process flexibility sometimes and may generate changes in equipment behavior that must be tracked and analyzed for possible re-training of the tool.

11.3.5 Case Study Conclusions

The following were the most relevant conclusions of this case study:

Concerning the prediction of anomalies in the operation of equipment with complex operation regime and poor monitoring of their condition through ANN AI-AR DM models:

- It was agreed that it is possible to model the behavior of the pumps and detect anomalies in their operation by deviations from the prediction. For instance, it could be appreciated how a pump that suffered a catastrophic failure after its first overhaul, worked with anomalous behavior throughout its second period of operation before the failure. Error accumulation was predicting this abnormal behavior. Another clear appreciation was the punctual accumulations of error detected when the asset was reaching certain operating regimes.

Concerning the guidelines for the practical implementation of the models in business systems:

- It was agreed that the systems are currently registering enough data to proceed with the implementation of this type of tools, and that it would be much more convenient to have all variables in a digital format and integrated in the same information system, where the ANN algorithm is finally implemented. The implementation of the algorithm can be done by asset, for greater precision, without a significant increase in costs.

Concerning the possibility to obtain information to build a business case:

- It was agreed that the integrated ANN-DM tool could avoid premature deterioration of equipment, verify the quality of its major maintenance intervention, ensure its operation at maximum energy efficiency, improve alarms and current interlocks for the operation of equipment, generate a new technical services to offer to potential clients, provide thoroughly understanding of the behavior of pumps throughout their life cycle, adjust the periods of completion of major maintenance and estimate the health of the asset better (comparing with current available techniques).

Concerning the identification and measurement of the asset´s loss of performance when anomalies are detected:

- It was agreed that the analysis of performance in similar operating regimes could be done properly. For instance, the measurement of the performance loss of the pump with the catastrophic failure, after its first overhaul could be done showing that the pump always operated with 4% to 6% lower performance. After the second overhaul, after the catastrophic failure, the pump perfectly recovered the initial performance.

In general terms, the implementation of the model has been considered a positive learning experience that can now be deployed for the entire fleet of similar assets, once the business case is approved. Another important aspect of this project implementation was to estimate the time schedule for this type of project. Figure 11.7 presents the percentage of time per process module accomplished when studying the first pump. It is important to notice the following:

- The input data processing time resulted in being a 50% of the total project time in this initial implementation, because of the dispersion, different formats and quality of the data processed. The total time when replicating the study to a similar asset

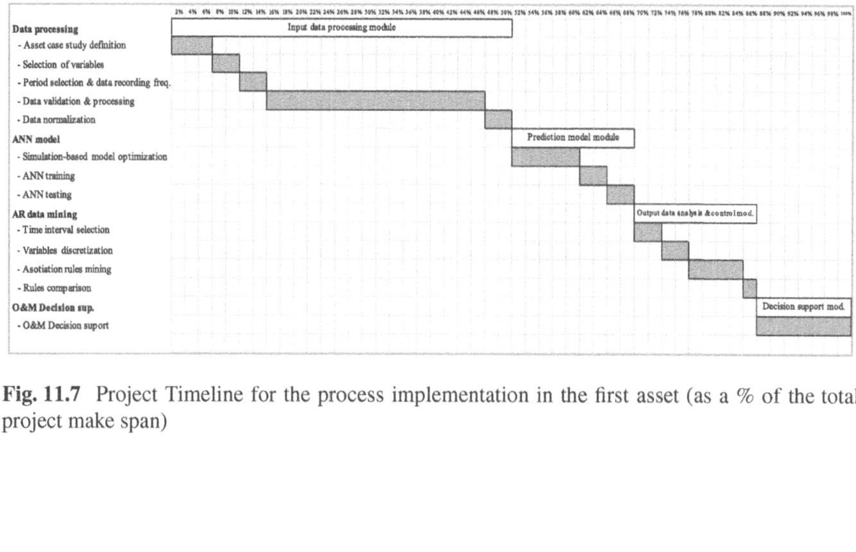

Fig. 11.7 Project Timeline for the process implementation in the first asset (as a % of the total project make span)

was reduced by 30%. Mainly impacting the input data processing module (20%-time reduction), and the same impact in the next two modules (5%-time reduction each).

- Time for data processing resulted longer than for prediction and data analysis modules together. This was a shock to the team involved in the project, even if the processing is in line with the one indicated by the Cross Industry Standard Process for Data Mining (CRISP-DM), which considers it to be around the 70% of the total processing time [58]. The use of the AI-DM tools for the problem analysis was expected to be more time consuming than the data processing according to the complexity of the tools that were going to be used. However, two intermediate processes were efficient and less dependent on circumstances out of the team control.

Being the current approach based on data, the main success factor of the methodology is data quality; however, this aspect can represent a major limitation for implementing it, since a substantial amount of data may not be immediately available in every company. At first, as noticed before, the procedure may appear time-consuming, but when it is included in the organizational routine, the obstacle characterized by the time for data validation and processing can be overcame. In addition, the application of ANN and AR mining requires an expertise in data understanding and modelling a positive mindset towards changing the current habits in terms of decision making in the O&M field.

References

1. Polo FAO, Bermejo JF, Fernández JFG, Márquez AC (2015) Failure mode prediction and energy forecasting of PV plants to assist dynamic maintenance tasks by ANN based models. Renew Energy 81:227–238
2. Mellit A, Benghanem M, Kalogirou SA (2007) Modeling and simulation of a stand-alone photovoltaic system using an adaptive artificial neural network: proposition for a new sizing procedure. Renew Energy 32:285–313
3. Kusiak A, Verma A (2012) A data-mining approach to monitoring wind turbines. IEEE Trans Sustain Energy 3:150–157
4. Antonanzas J, Osorio N, Escobar R, Urraca R, Martinez-de-Pison FJ, Antonanzas-Torres F (2016) Review of photovoltaic power forecasting. Sol Energy 136:78–111
5. Voyant C, Notton G, Kalogirou S, Nivet ML, Paoli C, Motte F, Fouilloy A (2017) Machine learning methods for solar radiation forecasting: A review. Renew Energy 105:569–582
6. Li Z, Rahman SM, Vega R, Dong B (2016) A hierarchical approach using machine learning methods in solar photovoltaic energy production forecasting. Energies 9:55
7. Minemoto T, Nakada Y, Takahashi H, Takakura H (2009) Uniqueness verification of solar spectrum index of average photon energy for evaluating outdoor performance of photovoltaic modules. Sol Energy 83:1294–1299
8. Mellit A, Kalogirou SA (2008) Artificial intelligence techniques for photovoltaic applications: a review. Prog Energy Combust Sci 34:574–632
9. Kalogirou SA (2011) Artificial neural networks in renewable energy systems applications: a review. Renew Sustain Energy Rev 5:373–401
10. Kalogirou SA, Bojic M (2000) Artificial neural networks for the prediction of the energy consumption of a passive solar building. Energy 25:479–491

11. Renno C, Petito F, Gatto A (2016) Artificial neural network models for predicting the solar radiation as input of a concentrating photovoltaic system. Energy Convers Manag 106:999–1012
12. Polson NG, Sokolov VO (2017) Deep learning for short-term traffic flow prediction. Transp Res Part C Emerg Technol 79:1–17
13. Wang HZ, Li GQ, Wang GB, Peng JC, Jiang H, Liu YT (2017) Deep learning-*based* ensemble approach for probabilistic wind power forecasting. Appl Energy 188:56–70
14. Li C, Bai Y, Zeng B (2016) Deep Feature Learning Architectures for Daily Reservoir Inflow Forecasting. Water Resour Manag 30:5145–5161
15. Bishop CM (1995) Neural networks for pattern recognition. Oxford University Press, Oxford
16. Boser BE, Guyon IM, Vapnik VN (1992) A training algorithm for optimal margin classifiers. In: Proceedings of the fifth annual workshop on computational learning theory. ACM, pp 144–152
17. Li Q, Meng Q, Cai J, Yoshino H, Mochida A (2009) Predicting hourly cooling load in the building: a comparison of support vector machine and different artificial neural networks. Energy Convers Manag 50:90–96
18. Zhang H, Chen L, Qu Y, Zhao G, Guo Z (2014) Support vector regression based on grid-search method for short-term wind power forecasting. J Appl Math 2014:1–11
19. Elyan E, Gaber MM (2017) A genetic algorithm approach to optimising random forests applied to class engineered data. Inf Sci 384:220–234
20. Shi J, Lee WJ, Liu Y, Yang Y, Wang P (2012) Forecasting power output of photovoltaic systems based on weather classification and support vector machines. IEEE Trans Ind Appl 48:1064–1069
21. Lin Y, Kruger U, Zhang J, Wang Q, Lamont L, El Chaar L (2015) Seasonal analysis and prediction of wind energy using random forests and ARX model structures. IEEE Trans Control Syst Technol 23:1994–2002
22. Moutis P, Skarvelis-Kazakos S, Brucoli M (2016) Decision tree aided planning and energy balancing of planned community microgrids. Appl Energy 161:197–205
23. Ren L, Hartnett M (2017) Prediction of surface currents using high frequency CODAR data and decision tree at a marine renewable energy test site. Energy Proc 107:345–350
24. Niu G, Yang BS, Pecht M (2010) 2010, Development of an optimized condition-based maintenance system by data fusion and reliability-centered maintenance. Reliab Eng Syst Saf 95:786–796
25. Jardine A, Lin D, Banjevic D (2006) A review on machinery diagnostics and prognostics implementing condition-based maintenance. Mech Syst Signal Process 20:1483–1510
26. Campos J (2009) 2009, Development in the application of ICT in condition monitoring and maintenance. Comput Ind 69:1–20
27. Lee J, Ghaffari M, Elmeligy S (2011) Self-maintenance and engineering immune systems: towards smarter machines and manufacturing systems. Annu Rev Control 35:111–122
28. Vachtsevanos G, Lewis F, Roemer M, Hess A, Wu B (2006) Intelligent fault diagnosis and prognosis for engineering systems. Wiley, Hoboken
29. Zio E (2009) Review reliabilty engineering: old problems and new challenges. Reliab Eng Syst Saf 94:125–141
30. United States Army. *ADS-79D-HDBK—Aeronautical Design Standard Handbook for Condition Based Maintenance Systems for US Army Aircraft*; US Army Aviation and Missile Research, Development and Engineering Center, Redstone Arsenal, Alabama,USA 2013.
31. Willis HL (2004) Power distribution planning reference book. CRC Press, Florida
32. González-Prida V, Guillén A, Gómez J, Crespo A, De la Fuente A (2019) An approach to quantify value provided by an engineered asset according to the ISO 5500x series of standards. In: Asset intelligence through integration and interoperability and contemporary vibration engineering technologies. Springer, Cham (2019), pp 189–196
33. Basheer IA, Hajmeer M (2000) Artificial neural networks: fundamentals, computing, design, and application. J Microbiol Methods 43:3–31

34. Zhang G, Patuwo BE, Hu MY (1998) Forecasting with artificial neural networks: the state of the art. Int J Forecast 14:35–62
35. Curry B, Morgan P, Beynon M (2000) Neural networks and flexible approximations. IMA J Manag Math 11:19–35
36. Kuo C (2010) Cost efficiency estimations and the equity returns for the US public solar energy firms in 1990–2008. IMA J Manag Math 22:307–321
37. Gibert K, Swayne D, Yang W, Voinov A, Rizzoli A, Filatova T (2010) Choosing the right data mining technique: classification of methods and intelligent recommendation. In: Modelling for environment's sake: proceedings of the 5th biennial conference of the international environmental modelling and software society, iEMSs 2010, Ottawa, Canada, July 2010, vol 3, pp 2457–2464
38. Buddhakulsomsiri J, Siradeghyan Y, Zakarian A, Li X (2006) Association rule-generation algorithm for mining automotive warranty data. Int J Prod Res 44:2749–2770
39. Agrawal R, Imieliński T, Swami A (1993) Mining association rules between sets of items in large databases. ACM SIGMOD Rec 22:207–216
40. Chen WC, Tseng SS, Wang CY (2005) A novel manufacturing defect detection method using association rule mining techniques. Expert Syst Appl 29:807–815
41. Kamsu-Foguem B, Rigal F, Mauget F (2013) Mining association rules for the quality improvement of the production process. Expert Syst Appl 40:1034–1045
42. Martínez-De-Pisón FJ, Sanz A, Martínez-De-Pisón E, Jiménez E, Conti D (2012) Mining association rules from time series to explain failures in a hot-dip galvanizing steel line. Comput Ind Eng 63:22–36
43. Sammouri W, Come E, Oukhellou L, Aknin P, Fonlladosa C-E (2013) Floating train data systems for preventive maintenance: a data mining approach. In: Proceedings of the 2013 international conference on industrial engineering and systems management (IESM), Rabat, Morocco, 28–30 October 2013; pp 1–7
44. Bastos P, Lopes ID, Pires L (2012) A maintenance prediction system using data mining techniques. World Congr Eng 3:1448–1453
45. Djatna T, Alitu IM (2015) An application of association rule mining in total productive maintenance strategy: an analysis and modelling in wooden door manufacturing industry. Proc Manuf 4:336–343
46. Antomarioni S, Bevilacqua M, Potena D, Diamantini C (2019) Defining a data-driven maintenance policy: an application to an oil refinery plant. Int J Qual Reliab Manag 36:77–97
47. Antomarioni S, Pisacane O, Potena D, Bevilacqua M, Ciarapica FE, Diamantini C (2019) A predictive association rule-based maintenance policy to minimize the probability of breakages: application to an oil refinery. Int J Adv Manuf Technol 2019:1–15
48. Nisi M, Renga D, Apiletti D, Giordano D, Huang T, Zhang Y, Mellia M, Baralis E (2019) Transparently mining data from a medium-voltage distribution network: a prognostic-diagnostic analysis. In: CEUR workshop proceedings, vol 2322, (CEUR-WS, 2019)
49. Hu L, Liu K-Y, Diao Y, Meng X, Sheng W (2016) Operational reliability evaluation method based on big data technology. In: Proceedings of the 2016 international conference on cyber-enabled distributed computing and knowledge discovery (CyberC), Chengdu, China, 13–15 October 2016, pp 341–344
50. Han J, Cheng H, Xin D, Yan X (2007) Frequent pattern mining: current status and future directions. In: Data mining and knowledge discovery, vol 15, No. 1, Springer Science + Business Media, Germany, pp 55–86
51. Fayyad UM, Irani KB (1993) Multi-interval discretization of continuous-valued attributes for classification learning. In: Proceedings of the 13th international joint conference on artificial intelligence 1022–1027. Chambery, France — August 28 - September 03, 1993.
52. Mata J, Alvarez JL, Riquelme JC (2003) An evolutionary algorithm to discover numeric association rules. In: Proceedings of the 2002 ACM symposium on applied computing, Madrid, Spain, 11–14 March 2003
53. Mirebrahim SH, Shokoohi-Yekta M, Kurup U, Welfonder T, Shah M (2017) A clustering-based rule-mining approach for monitoring long-term energy use and understanding system behavior.

In: Proceedings of the 4th ACM international conference on systems for energy-efficient built environments, Delft, The Netherlands, 8–9 November 2017, pp 5:1–5:9

54. Jović A, Brkić K, Bogunović N (2014) An overview of free software tools for general data mining. In: Proceedings of the 2014 37th international convention on information and communication technology, electronics and microelectronics, MIPRO 2014-Proceedings, Opatija, Croatia, 26–30 May 2014
55. Announcing the Standard for Integration Definition For Function Modeling (IDEF0). Draft Fed. Inf. Proc. Stand. Publi. 1993:183
56. Crespo Márquez A, De la Fuente A, Antomarioni S (2019) A process to implement artificial neural network and association rules techniques to improve assets performance and energy efficiency. Energies 2019 12(3454), 1–25. https://doi.org/10.3390/en12183454
57. Theodoridis S (2015) Probability and stochastic processes. In: Chapter 2 - Probability and stochastic processes, machine learning. Academic, pp 9–51
58. Chapman P, Clinton J, Kerber R, Khabaza T, Reinartz T, Shearer C, Wirth R (1999) The CRISP-DM user guide. In: Proceedings of the 4th CRISP-DM SIG workshop in Brussels in March, Brussels, Belgium

Part VI
Emerging Data-Driven Processes for Asset Investment Planning (AIP)

Chapter 12
Asset Health Indexing and Life Cycle Costing

12.1 Introduction

An Asset Health Index (AHI) is an asset score, which is designed, in some way, to reflect or characterize the asset's condition and thus, its performance in terms of fulfilling the role established by the organization [1]. AHI represent a practical method to quantify the general health of a complex asset. For simple assessments, CBM technologies can precisely estimate the status of a specific asset with defined and specifics failure modes. However, most of these assets are composed of multiple subsystems, and each subsystem can be characterized by multiple modes of degradation and failure. From a pure theoretical perspective, every failure mode of every item that composes a system can be modelled and estimated. In some cases, it may be considered that an asset has reached the end of its useful life, when several subsystems have reached a state of deterioration that prevents the continuity of service required by the business [2]. This calculation can be complex and cause a significant investment in time and resources. It is in the case of complex systems where the health index, based on the results of operational observations, field inspections and laboratory tests, produces a single objective and quantitative indicator. It may be used as a tool to manage assets, to identify capital investment needs and maintenance programs, allowing [3–5]: (1) Compare the health of equipment located in similar technical locations, to study possible premature deterioration and optimize operation plans and/or asset maintenance if necessary; (2) Communicate more accurately with manufacturers/builders, to understand the behavior of assets of different manufacturers/builders in specific technical locations; and (3) Support decision-making processes in future investments in assets, or in extension of the life of these [6].

Asset Health Indicators are widely used in supporting maintenance and replacement strategies based on asset condition and performance in some countries, to justify asset replacement schemes to the regulators [7–9].

A proper design of a health index should meet the following requirements [10]:

- The index must be indicative of the suitability of the asset to provide continuity to the service and representative of the general health of the asset.

A. Crespo Márquez, *Digital Maintenance Management*, Springer Series in Reliability Engineering, https://doi.org/10.1007/978-3-030-97660-6_12

- The index should contain objective and verifiable measures of the condition of the asset, instead of subjective observations.
- The index must be understandable and easily interpreted.

Several methods and models fulfilling these requirements have been reviewed, for instance, the ones by Kinetrics [3, 4], DNV GL [11], Terna [12] and GB DNO [7].

Although most of these models build a streamlined approach to introduce different influent factors to estimate the lifetime expectation/remaining useful life of an asset, several drawbacks are still present in their model formulation:

(i) The AHI procedure seems not to be properly robust from the scientific perspective, original models are built mostly by practitioners in specific sectors with very specific assets.
(ii) Many influent factors are evaluated based on assumptions that are never discussed (e.g., ranges of numerical values are given as scales for different factors while it is almost unclear what is the basis to define such ranges);
(iii) The procedure proposed is mainly presented in its development and never demonstrated completely with, at least, some case-based reasoning or at the minimum a complete industrial case which would enable a proper validation of the AHI model proposed.

There are approaches in the relevant literature to identify asset health [13] used mainly in CBM applications based on dynamic health assessment, but the concept is different from the one used in this paper, now the health assessment allows comparison and decision-making among different assets.

To overcome these weakness points, in this paper the methodology adopted to model the AHI is only loosely based on the OFGEM Network Asset Indices Methodology [7] (a similar approach as in the example previously presented in [14]). This method is selected because it is considered simple for simulation model building purposes and very practical in its implementation, if a more robust scientific design of the model format is reached.

More precisely, the method [7] requires: (1) The identification of the asset, which includes the category of the equipment under study, the current age, the expected life, the name of the manufacturer/builder, the model of the equipment and the location of the installation; (2) The operation and maintenance data recorded during a certain period of time; and (3) The condition of the equipment, that is, the results of the analyses performed on the equipment in site, results of readings of physical variables, results of visual inspections, etc.

The health index model adopted in this paper contains values between 1 and 10, thus being able to compare health between different types of assets. There are other indices that go from 0 to 1 and others that range from 1 to 100. In any case, they all have the same functionality, normalize the health of different assets to be able to compare them with each other.

12.2 AHI Modelling Methodology

The application procedure for calculating the health index is based on 6 consecutive steps, in which, starting from a design life associated with an equipment's category, a current health index is reached. For this, a series of factors related to the location, operation and condition of the asset are considered. It is presented in the following Fig. 12.1, the model, with the 6 steps for calculating the health index of an asset. For a precise description of the methodology and formulation of the AHI the reader is addressed to [15].

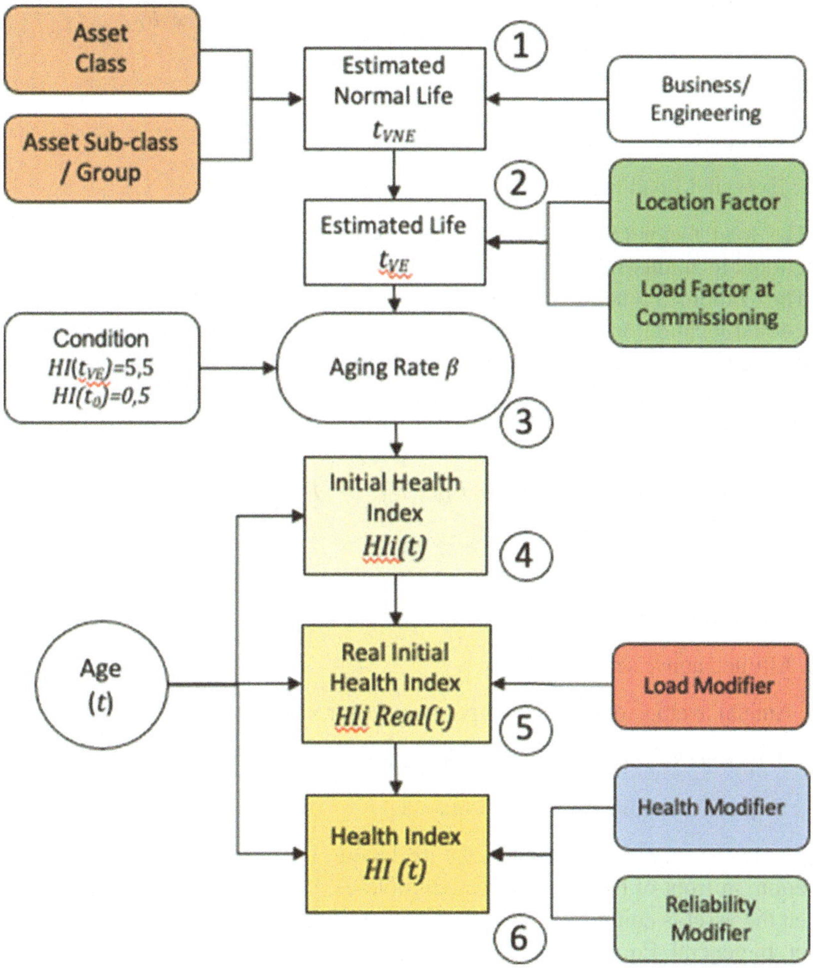

Fig. 12.1 Procedure to calculate the AHI

A synthesis of formulation is as follows: In step 1 a design theoretical life for every asset depending on the equipment category is defined. Once identified manufacturing/built data, model and technical design specifications of an asset, its design life can be adapted by the owner according to accumulated experience and the information provided by different manufacturers and builders. This estimated owner life will be considered for accounting purposes and to measure asset depreciation and asset book value over time.

In step 2, the estimated owner life can then be adjusted according to the characteristics of the asset location and loading.

$$Estimatedlife = t_{EL} = \frac{Designlife(books)}{F_{FL} \cdot F_{EL}} = \frac{t_{DL}}{F_{FL} \times F_{EL}} \quad (12.1)$$

where

F_{FL}: Combined functional location factor.

F_{EL}: Expected Load factor.

The load factor (F_C), as well as the location factor (F_E), is inherent in the functional location of the asset.

The Combined functional location factor (F_{FL}) is based on the impact of the asset functional location on its operation and maintenance. The specific location factors considered are proposed by GB DNO group methodology [7] and must be adapted depending on the specific industry and on the specific asset [16].

With:

$$F_{FL} = \max(F_{DC}, F_A, F_T, F_{AT}, F_{PS}) \quad (12.2)$$

And

F_{DC}: Distance to the coast factor.

F_A: Altitude above sea level factor.

F_T: Annual average of outside temperature factor.

F_{AT}: Exposure to corrosive atmosphere factor.

F_{PS}: Exposure to dust in suspension factor.

The load factor (F_{EL}) measures the load request that is made on the asset in that location, in front of the maximum admissible load. Normally, this data is provided when the asset is commissioned, and it is part of the technical specifications of the asset. In general, Eq. 12.3 is used:

$$F_{EL} = \frac{Loadundernormaloperatingconditions}{Maximumpermissibleload} \quad (12.3)$$

A fundamental hypothesis of the methodology is that the irreversible degradation of an asset follows an exponential behavior with respect to its age, and in step number 3, the aging rate (β) of the asset is determined by the natural logarithm of the quotient between the asset health index when new (H_{new}) and the asset health index when reaching its expected life ($H_{estimatedlife}$). The equation for its calculation is the following, used in step 3:

$$\beta = \frac{\ln \frac{H_{new}}{H_{estimatedlife}}}{Estimatedlife} = \frac{\ln \frac{0,5}{5,5}}{t_{EL}} \tag{12.4}$$

With:

Estimated life: Time calculated in (12.1).

HI_{new}: Health index for a new asset.

$HI_{estimatedlife}$: Health index for the asset expected life time.

Then, in step 4, the initial health index (HIi) is considered as a dimensionless number between 0.5 and 10, with an exponential behavior with respect to the age "t" of the asset, which is characterized by the aging rate as follows (Fig. 12.2):

$$HIi_t = HI_{new} \cdot e^{\beta \cdot t} \tag{12.5}$$

The health index (*HI*) is the result of adjusting the initial health index, using load, health, and reliability modifiers. In a first step, the initial health index (*HIi*) of an asset is modified to obtain what we call the real initial health index (Real HIi) in step 5, considering the load modifier registered for the current age ($M_L(t)$), using the following equation:

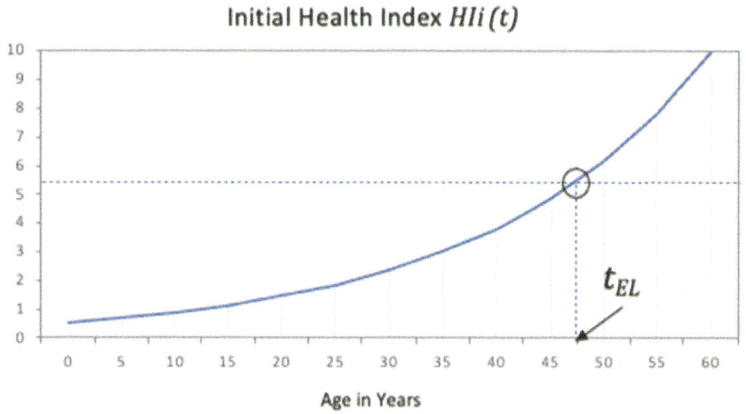

Fig. 12.2 Graph of the initial asset health index (HIi)

$$HI_i Real(t) = HI_{new} \cdot e^{\frac{\beta \cdot t}{M_L(t)}} \tag{12.6}$$

where the load modifier is the quotient between the load factor existing at an instant $(F_{RL}(t))$ and the expected load factor (F_{EL}).

$$M_L(t) = F_{EL}/F_{RL}(t) \tag{12.7}$$

The load modifier is a health modifier of the asset, which is considered in this initial phase since it is very likely that in many assets the load recorded during each asset age will be significantly different to the one initially planned for the functional location The introduction of HIi Real then allows the current asset degradation to be adjusted to compare with the anticipated degradation for the functional location.

Finally, in step 6, the health index of the asset is determined by its operating conditions and reliability conditions at the time of the evaluation. To determine the health index, the following equation is used:

$$HI(t) = HI_i Real(t)^{e^{(MH(t)+MR(t))}} \tag{12.8}$$

where

$MH(t)$: is the asset's health modifier (condition and operation).

$MR(t)$: is the asset's reliability modifier.

For the evaluation of the health modifier (*MH*) that appears in this last equation, the different variables that can be measured and quantified for each asset sub-category are considered, and that, being independent in their impact on health, add information about it. For the reliability modifier, depending on the sub-category of asset, model and manufacturer, tables can be prepared with the value of this parameter.

In general, the equations to obtain the value of the health modifier (*MS*) and the reliability modifier (*MF*) will be the following:

$$MH(t) = \sum_{j=1}^{j=n} \gamma_j \cdot MH_j(t) \tag{12.9}$$

With

$j = 1 \ldots n$ is an index used for different health modifiers.

γ_j: is the weight assigned to the health modifier j in the model.

$MH_j(t)$: is the health modifier at time t, age of the asset.

And

$$MR(t) = \sum_{k=1}^{k=m} \gamma_k \cdot MR_k(t) \tag{12.10}$$

With

$k = 1 \ldots m$ is an index used for different reliability modifiers.

γ_k: is the weight assigned to the reliability modifier k in the model.

$MR_k(t)$: is the reliability modifier at time t, age of the team.

The determination of the weight assigned to each modifier must be done relative to the rest of the modifiers and assuming a maximum possible impact of the set of modifiers in their worst condition. This is achieved considering the following restrictions:

The sum of the totality of the modifier weights will be equal to the so-called maximum impact rate γ (for this rate, all modifiers always take the value 1).

$$\sum_{j=1}^{j=n} \gamma_j + \sum_{k=1}^{k=m} \gamma_k = \gamma \tag{12.11}$$

The value of γ is obtained by forcing the IHi to be equal to 10 (maximum limit of the *HI* of the asset in the model) upon reaching the estimated normal life (t_{EL}) of the asset:

$$HI_i(t_{EL})^{e^{\gamma}} = 10, then(5, 5)^{e^{\gamma}} = 10$$

And therefore:

$$\gamma = Ln(Ln(10)/Ln(5.5)) = 0.301$$

The effect that the modifiers of the asset's health have in the calculation of the health index can be seen as an example in Fig. 12.3. Where the $HIi(t)$ is compared to the $RealHIi(t)$ and the $HI(t)$. This Figure considers the possibility that the asset's health index improves with respect to the forecast (by reducing the load compared to the forecast). The figure includes the effect of the Overhaul of the pump on the mentioned indices.

MH and *MR* will be standardized values within the interval [0, 1]. Finally, the actual health index of a system can be obtained and represented as in Fig. 12.4, where each asset conforming the system may have a different degradation speed. Asset initial condition can be restored after major maintenance intervention or assets renovations or replacement.

Table 12.1 provides a simple interpretation for an AHI, in this example and for an asset with a normal life expected of 50 years, recommendations for action are adjusted.

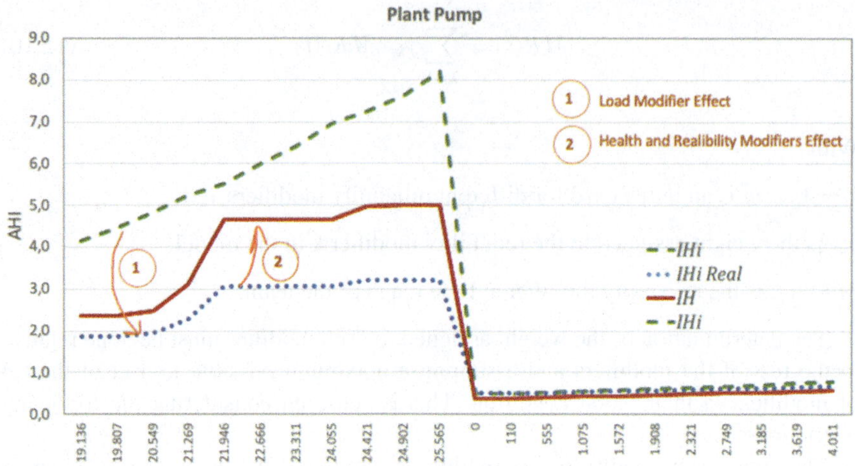

Fig. 12.3 Effect of modifiers on the initial health index of assets

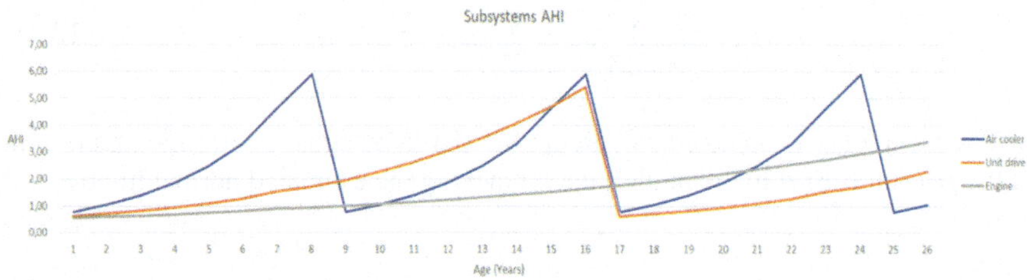

Fig. 12.4 Health index for different assets of a system (one colour per asset)

Table 12.1 Asset health index and expected lifetime

AHI	Condition	Expected lifetime	Requirements
0.5–4	Very good	More than 15 years	Normal maintenance
4–5.5	Good	More than 10 years	Normal maintenance
5.5–7	Fair	From 3–10 years	Increase diagnostic testing, possible replacement depending on criticality
7–8	Poor	Less than 3 years	Start planning process to replace
8–10	Very poor	Near to the end of life	Immediately assess risk; replace or rebuild based on assessment

12.3 Life Cycle Costing Modelling Methodology

The application procedure for calculating life cycle cost is based on a total expenditure (TOTEX) approach, which takes a holistic view to model the cost of the asset. The

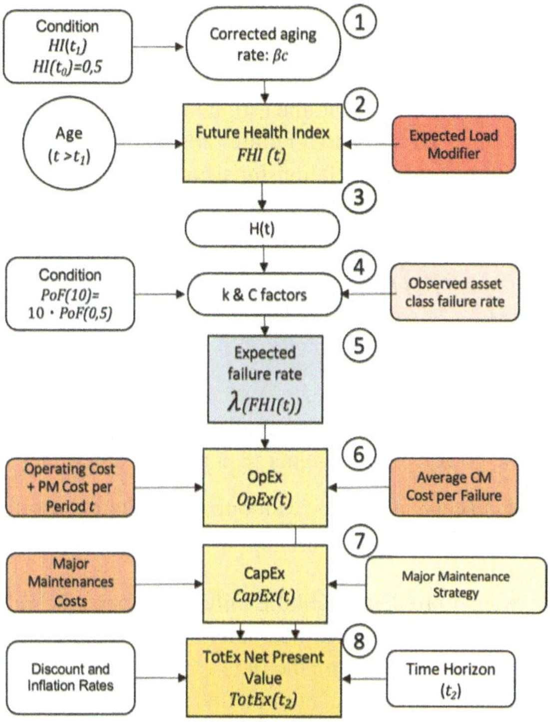

Fig. 12.5 Procedure to Calculate the TOTEX of the asset until the end of its live Cycle

model is based on 8 consecutive steps, departing from a point where the asset health index has been estimated using previous methodology (t_1). In Fig. 12.5, the model with the 8 steps to estimate the NPV of the TOTEX of an asset, is presented.

At the starting point, the asset's health index at that instant is calculated $HI(t1)$. Then, through Step 1 the corrected aging rate is obtained. This aging rate considers the set of circumstances that took place from the asset's start-up (or the last major maintenance that took place at time to) up to the current point t_1, using Eq. 12.12:

$$\beta c = \frac{\ln \frac{AHI_{t1}}{HI_{new}}}{t_1} \tag{12.12}$$

Step 2 projects the health index expected in the future up to the planning horizon, that is, in the interval (t_1, t_2) according to Eq. 12.13.

$$FHI_t = HI_{new} \cdot e^{\beta c \cdot t}, \quad \text{with} t_1 < t \leq t_2 \tag{12.13}$$

Step 3 is for the estimation of the function $H(t)$, which is a new function that is equal to:

$$H(t) = H_t = \begin{cases} HI(t), si HI(t) \geq 4 \\ 4, Otherwise \end{cases} \tag{12.14}$$

Step 4 is devoted to the estimation of factors C and K. In fact, the methodology links the *FHI* value to a failure rate (λ) of the asset. For instance, the methodology [6] formulates this relationship using the first three terms of the Taylor series for the exponential function as in Eq. 12.15:

$$\lambda(FHI(t)) = K \cdot \left[1 + (C \cdot H_t) + \frac{(C \cdot H_t)^2}{2!} + \frac{(C \cdot H_t)^3}{3!}\right] \tag{12.15}$$

where

C: Constant value, equal for all assets, (in [6] C value is selected such that the λ for $HI = 10$ is ten times higher than the λ of a new element.

K: Constant value to be calibrated, considering asset observed failure frequency per annum, the *HI* distribution of asset population and the volume of assets within the population.

Step 5 projects the future failure rate once the values of $H(t)$, K and C are estimated, according to Eq. 12.5, and Step 6 obtains the *OpEx* projection for a future period t by multiplying the future failure rate by the unit cost of failures, and to that value add the equipment operation and maintenance costs, as in Eq. 12.16.

$$OpEx(t) = CO(t) + \lambda(t) * CM + CMP(t) \tag{12.16}$$

Future CAPEX projection is obtained in Step 7 of the process, by identifying the time periods where the major maintenance will take place, based on the strategy determined by the organization to set them (based on hours of operation, asset health, or another strategy selected), as in Eq. 12.17:

$$CapEx(t) = CMM(t) * RMM(t, HI(t)) \tag{12.17}$$

where $RMM(t)$ is a Boolean function that determines whether major maintenance or "overhaul" (MM) is performed in a certain period (t), based on a certain strategy (for example, based on t or on $HI(t)$). $CMM(t)$, is the cost of the major maintenance to be carried out in period t. The cost of major maintenance may vary as the age of the equipment varies.

$$RMM(t) = \begin{cases} 1, & MM \ released \ at \ period \ t \\ 0, & Otherwise \end{cases} \tag{12.18}$$

In the last step of the procedure, step 8, the NPV of the future TotEx is calculated as in Eq. 12.19.

$$TotEx(t) = OpEx(t) + CapEx(t) \tag{12.19}$$

And finally, the next present value of the selected strategy is obtained in Eq. 12.20.

$$VANTotEx(TotEx(t), i, K, t1, t2) = \sum_{t=t1}^{t=t2} \frac{TotEx(t) \cdot (1+i)^{t-t1}}{(1+k)^{t-t1}} \tag{12.20}$$

12.4 Case Study. Lng Regassification Plant

The model has been used to develop a plan for capital investment (CapEx), operations and maintenance (OpEx) Expenditure to support the associated strategic decision-making processes in a Company (Regasification Terminal) for the period 2019–2035. The goal was to use the model to develop a profile of expected expenses associated with a total of 54 assets of a Regasification Terminal (the "Company"). The company decided to follow the approach established in the guides prepared for this purpose by the Institute of Assets Management (IAM) [17]. Consisting of the following steps:

1. Evaluation of the criticality of the Assets.
2. Assessment of the assets' health, both current and projected. Obtaining the health index to the condition of the assets, in order to obtain the probability of failure of each asset until the year 2035, based on the current failure probabilities and the evolution of the health of said assets throughout the period under study.
3. Forecast of the evolution of costs or economic risks throughout the life cycle of the Assets under study, considering the results of the previous point and the major revisions and major maintenance planned.
4. Study of the different strategies for major maintenance release, again applying the methodology used in points 2 and 3 above, to facilitate technical–economic decision making in relation to major revisions or the Equipment renewal, with a horizon of 2019–2035.
5. Making a better estimate, as of March 2019, of the expected expense profile for the assets, during the 2019–2035 period.

For this study, historical data recorded by the company, regarding O & M of these equipment since its commissioning (2009) has been available to the project review team.

12.4.1 Asset Expense Profile Based on Its Health Index

Keeping the asset's probability of failure within an acceptable range, requires preventive maintenance and major maintenance (also called "overhauls" of the equipment).

After the overhaul the asset is practically in a situation of "as new". Therefore, it is understood that by carrying out the overhaul when the $AHI = 5.5$, the failure rates will remain under control, as expected for the established maintenance plans. If this major maintenance is postponed, the frequency of failures will grow, and so will the increase in the needs for corrections and their costs, or the needs for preventive activities to retain reliability within a desired range, and their cost, decreasing efficiency in assets management.

Major maintenance activities schedules were originally recommended by the manufacturers and builders of each piece of equipment, according to functional location and operating conditions (and are often taken as generic or standard). However, after ten years of plant operation, the actual use of the asset or even its location, environmental conditions, etc., may differ from those that the manufacturer/builder had assumed in their recommendations. Of course, this may affect original failure rates considered and subsequently financial statements.

In this context, the analysis attempted to determine the multiplying factors that apply to the failure rate of each asset, because of aging (AHI), and the associated cost of maintenance, which is affected by these new failure rates. On the other hand, a new strategy was proposed for major maintenance scheduling, to avoid out-of-control increase in the failure rates. More precisely, the strategies that were analyzed were the following:

- Strategy 1. Total expenditure on the asset (TotEx) is calculated executing its maintenance according to the manufacturer's recommendations (as in Fig. 12.6), without incorporating substantive changes to the original maintenance methodology (desired age-based overhauls).
- Strategy 2. TotEx is estimated assuming a standard OpEx but adjusting the overhauls schedule (CapEx) in order to maintain a controlled level of reliability (as in Fig. 12.7), avoiding the increase in the failure rate beyond the standard values considered (i.e. health index-based overhauls, with $AHI \leq 5.5$).

In Figs. 12.8 and 12.9, projection of AHI values under both strategies are presented. Assets included within the named: KO Drum system of the plant are included. AHI projections and then translated to numbers in Table 12.2, for two

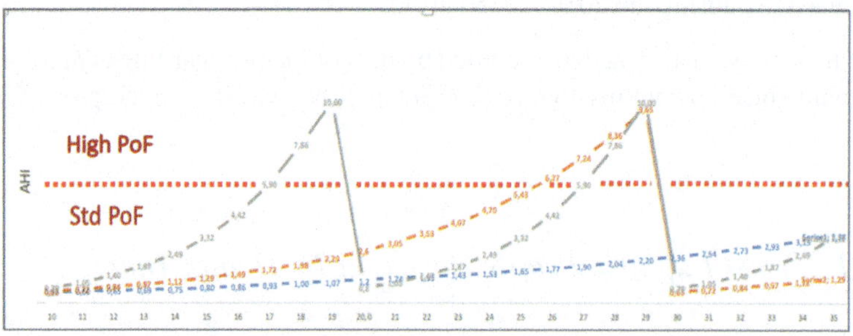

Fig. 12.6 Sample projection of AHI values for strategy 1

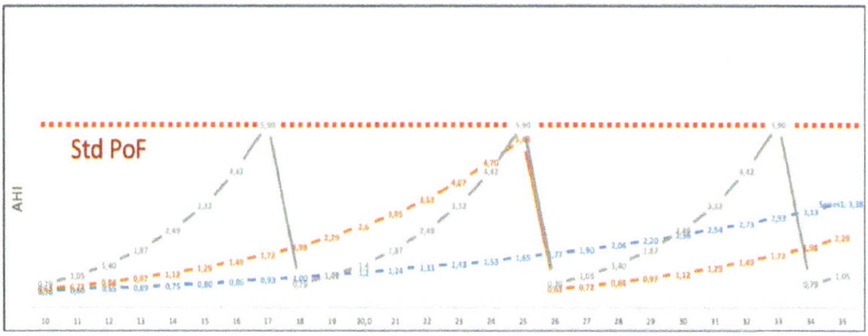

Fig. 12.7 Sample projection of AHI values for strategy 2

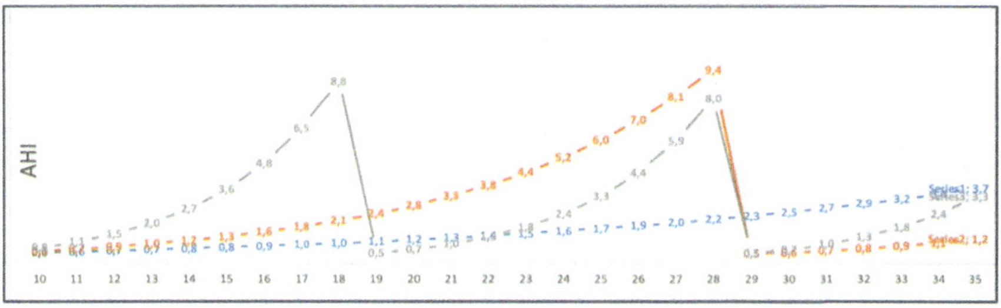

Fig. 12.8 Projection of AHI for Strategy 1 since 2019 and until 2035. KO Drum system

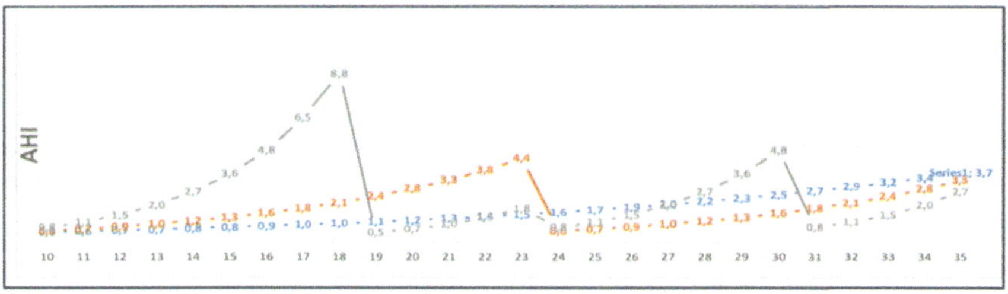

Fig. 12.9 Projection of AHI for Strategy 2 since 2019 and until 2035. KO Drum system

Systems KO Drum and Fender (notice that some of the systems may consist of only one asset). Once the list of expenses and investments is projected, a financial calculation is carried out to compare both strategies. In this calculation inflation is considered but assets depreciation is not. So, CapEx only includes expected reinvestments until the year 2035.

Table 12.2 Annual CapEx (without depreciation) & Opex for both strategies, and two assets (Ko drum and Fender ending 2035. Last two columns are for TotEx & VAN (Net Present Value) of Strategy 2 (the one that was selected)

Equipo	Descripción		Est	Capex /Opex	2018	2019	2020	2021	2022	2023	2024	2025	2026	2027	2028	2029	2030	2031	2032	2033	2034	2035	Totex EST 2 USD (2018-2035)	VAN EST2 (MM USD en 2018)
100-V-101	KO Drum	KO drum del muelle	EST 1	Capex	0,0	58,3	0,0	0,0	0,0	0,0	0,0	0,0	0,0	0,0	138,4	0,0	0,0	0,0	0,0	0,0	0,0	0,0	1.022.468	485.031
			EST 1	Opex	100,0	14,2	14,6	15,0	15,5	45,8	55,6	68,9	87,3	113,1	150,0	19,0	19,6	20,2	20,8	21,4	22,0	22,7		
			EST 2	Capex	0,0	58,3	0,0	0,0	0,0	119,4	0,0	0,0	0,0	0,0	0,0	0,0	80,8	0,0	0,0	0,0	0,0	0,0	738.573	382.706
			EST 2	Opex	100,0	14,2	14,6	15,0	15,5	45,8	16,4	16,9	17,4	17,9	18,5	19,0	61,8	20,2	20,8	21,4	22,0	22,7		
100-X-110C	Fender	Punto de apoyos y amortiguación Barco	EST 1	Capex	0,0	24,7	0,0	0,0	0,0	0,0	1,2	0,0	0,0	0,0	0,0	33,2	0,0	0,0	0,0	0,0	1,7	0,0	331.483	151.058
			EST 1	Opex	24,6	9,0	9,2	9,5	9,8	10,1	10,4	14,8	18,0	23,9	33,6	12,0	12,4	12,8	13,1	13,5	14,0	19,9		
			EST 2	Capex	0,0	24,7	0,0	0,0	0,0	0,0	1,2	0,0	0,0	30,0	0,0	1,4	0,0	0,0	0,0	0,0	1,7	0,0	315.385	145.996
			EST 2	Opex	24,6	9,0	9,2	9,5	9,8	10,1	10,4	14,8	18,0	22,6	11,7	12,0	12,4	12,8	13,1	13,5	19,3	23,5		

Table 12.3 CapEx (no depreciation included) and OpEx, over the years, for both strategies until 2035. VAN represent Net Present Value of Strategy 2 (Selected Strategy). Values in Thousands USD$. All assets under analysis

EST	Capex /Opex	2018	2019	2020	2021	2022	2023	2024	2025	2026	2027	2028	2029	2030	2031	2032	2033	2034	2035	VAN
EST 1	Capex	621	1.832	669	107	104	237	2.508	2.916	2.877	138	4.265	8.866	1.054	325	93	90	2.418	2.349	12.427 $
EST 1	Opex	6.911	7.161	7.036	7.126	7.372	7.766	8.108	8.304	8.565	8.927	9.445	9.658	9.426	9.618	9.830	10.166	10.520	10.808	66.667 $
EST 2	Capex	645	1.336	175	1.176	684	1.231	2.700	3.130	2.143	4.104	4.161	1.266	426	3.125	300	1.646	893	13	12.499 $
EST 2	Opex	6.757	7.132	6.893	7.103	7.362	7.723	7.942	8.025	8.271	8.557	8.854	9.127	9.333	9.657	9.867	10.173	10.456	10.786	65.467 $
EST 1	Totex	7.532	8.993	7.705	7.233	7.476	8.003	10.615	11.221	11.442	9.065	13.710	18.524	10.479	9.944	9.923	10.256	12.938	13.157	79.093 $
EST 2	Totex	7.402	8.469	7.068	8.279	8.046	8.954	10.643	11.155	10.414	12.661	13.015	10.393	9.759	12.782	10.167	11.819	11.349	10.799	77.966 $

12.4.2 Industrial Case Results and Findings

The analysis identified Strategy 2 as the most appropriate for the fulfillment of dependability targets of the terminal. This was the only specific strategy to ensure assets' reliability within standard limits in this type of industry. Additionally, and curiously, Strategy 2 scored the lowest updated TotEx value (77,966 M USD). The savings in higher reliability, are higher than the extra capital investment required in number of overhauls. Results obtained for the total 54 assets are included in Table 12.3.

References

1. De la Fuente A, Crespo A, Sola A, Guillén A, Gómez J, Amadi-Echendu JE (2021) Planning major overhaul and equipment renovation based on asset criticality and health index. In: Crespo Márquez A, Komljenovic D, Amadi-Echendu J (eds) 14th WCEAM proceedings, WCEAM 2019. Lecture notes in mechanical engineering. Springer, Cham. https://doi.org/10.1007/978-3-030-64228-0_8
2. Hjartarson T, Otal S (2006) Predicting future asset condition based on current health index and maintenance level. In: ESMO 2006–2006 IEEE 11th international conference on transmission and distribution construction, operation and live-line maintenance
3. Naderian A, Cress S, Piercy R, Wang F, Service J (2008) An approach to determine the health index of power transformers. In: Conference record of the IEEE international symposium on electrical insulation, 192–196. https://doi.org/10.1109/ELINSL.2008.4570308

4. Naderian A, Piercy R, Cress S, Service J, Wang F (2009) An approach to power transformer asset management using health index IEEE Electr Insul Mag 25(2):2. https://doi.org/10.1109/MEI.2009.4802595
5. Azmi A, Jasni J, Azis N, Kadir MZA (2017) Evolution of transformer health index in the form of mathematical equation. Renew Sustain Energy Rev 76(March):687–700. https://doi.org/10.1016/j.rser.2017.03.094
6. Silvestri L, Forcina A, Introna V, Santolamazza A, Cesarotti V (2020) Maintenance transformation through Industry 4.0 technologies: a systematic literature review. Computers in industry, vol 123. ISSN 0166-3615. https://doi.org/10.1016/j.compind.2020.103335
7. GB DNO groups. DNO common network asset indices methodology (2017). Health & criticality—version 1.1. https://www.ofgem.gov.uk/system/files/docs/2017/05/dno_common_network_asset_indices_methodology_v1.1.pdf
8. Australian Local Government Association (2015). National State of the assets report. https://alga.asn.au/national-state-of-the-assets-report-2015/
9. Federation of Canadian Municipalities (FCM) & other seven partner organizations (2020). The Canadian Infrastructure Report Card (CIRC) 2019 http://canadianinfrastructure.ca/downloads/canadian-infrastructure-report-card-2019.pdf
10. Hjartarson T, Otal S (2006) Predicting future asset condition based on current health index and maintenance level. In: ESMO 2006–2006 IEEE 11th international conference on transmission & distribution construction, operation and live-line maintenance
11. Vermeer M, Wetzer J, Wielen P, van der, de Haan E, de Meulemeester E (2015) Asset management decision support modeling, using a health and risk model. PowerTech IEEE Eindhoven 1–6. https://doi.org/10.1109/PTC.2015.7232556
12. Scatiggio F, Pompili M (2013) Health index: the TERNA's practical approach for transformers fleet management. IEEE Electr Insul Conf (EIC) 178–182
13. de Calle KL, Ferreiro S, Arnaiz A, Sierra B (2019) Dynamic condition monitoring method based on dimensionality reduction techniques for data-limited industrial environments. Computers in industry, vol 112, p 103114. ISSN 0166-3615. https://doi.org/10.1016/j.compind.2019.07.004
14. Crespo Márquez A, et al (2020) Defining asset health indicators (AHI) to support complex assets maintenance and replacement strategies. A generic procedure to assess assets deterioration. In: Crespo Márquez A, Macchi M, Parlikad A (eds) Value based and intelligent asset management. Springer, Cham. https://doi.org/10.1007/978-3-030-20704-5_4
15. Serra J, De la Fuente A, Crespo A, Sola A, Guillén A, Candón E, Martínez-Galán P (2019) A model for lifecycle cost calculation based on asset health index. In: International Conference on Smart Infrastructure and Construction 2019 (ICSIC), pp 91–98. ISBN 978-0-7277-6466-9. https://www.icevirtuallibrary.com/. https://doi.org/10.1680/icsic.64669.091
16. De la Fuente, A. Guillén, A. Crespo, A. Sola, J. Gómez, P. Moreu, V. Gonzalez-Prida (2018) Strategic view of an assets health index for making long-term decisions in different industries. In: Chapter 126, safety and reliability—safe societies in a changing world: proceedings of ESREL 2018, June 17–21, 2018, 1st edn. Trondheim, Norway, CRC Press. https://doi.org/10.1201/9781351174664
17. Institute of Assets Management (IAM): subjects 6 and 7: capital investment operation and maintenance decision-making. Version 1.0 (2016)